FINITE MATH
for
LIBERAL ARTS

FINITE MATH

for

LIBERAL ARTS

Dilip K. Datta

Department of Mathematics
University of Rhode Island

RHODE ISLAND DESKTOP
393 Biscuit City Road
Kingston, Rhode Island 02881

Finite Math for Liberal Arts

Second Edition

© 2000 Dilip K. Datta

ISBN 0-9638605-3-4

Cover design by Jeannette Woodall

Jeannette Woodall (b. 1921) of 26 Champlin Drive, Avondale, Rhode Island
02891 is a member of the Mystic Art Association. She received part of her training
in Lacoste Art School of Provence, France. Jeannette holds a B.F.A. degree from
the University of Rhode Island and is a member of the National Scholastic Honor
Society. Jeannette is a busy artist. In the spring of 1994, she set aside her brushes
to take a Finite Math course. Fortunately, the 73 year old artist student of math
found inspiration to do a piece of creative work. In the cover design she has tried
to capture the mood, the content and the meaning of the course set on the backdrop
of her favorite topic -- the Venn diagrams.

RHODE ISLAND DESKTOP
393 Biscuit City Road
Kingston, Rhode Island 02881

TABLE OF CONTENTS

DEPENDENCE CHART   vi

PREFACE   vii

ACKNOWLEDGMENT   ix

SUGGESTED READINGS   x

SUGGESTED LEARNING AIDS   x

I.   ORGANIZATION AND DISPLAY OF DATA
    1. Frequency distributions   1
    2. Relative frequency   7
    3. Circle graphs or pie charts   11
    4. Stem and Leaf Displays   16
    5. Grouped data   20
    6. Ranked data   27
  *7. Baseball and Basketball   31
       Chapter test   34

II. ANALYSIS OF DATA
    1. Masures of central tendency   36
    2. The mean of grouped data   46
    3. Measures of variation   49
    4. Correlation Coefficient   58
       Chapter test   65

III. SET THEORY
    1. Yes-No Questionnaire   67
    2. Sets and elements   76
    3. Subsets   84
    4. One-to-one correspondence   87
    5. Counting with more than Two Sets   89
       Chapter test   94

IV. COUNTING TECHNIQUES
   1. Tree diagrams   96
   2. The fundamental principle of counting   100
   3. Permutations   103
   4. Combinations   106
       Chapter test   111

V. PROBABILITY THEORY
   1. Probability measure   113
   2. Equiprobable measure   120
   3. Properties of Probability measure   125
   4. Odds and fair bet   131
   5. Conditional probability   135
       Chapter test   144

VI. BINOMIAL PROBABILITY
   1. Finite stochastic process   146
   2. Independent trials with two outcomes   155
   3. Table for binomial probabilities   160
   4. Expected number of successes   162
   5. Standard deviation of the binomial distribution   166
       Chapter test   171

VII.   BINOMIAL DISTRIBUTION AND THE NORMAL CURVE
   1. Binomial distribution   174
   2. The normal curve   180
   3. Estimating binomial probability by the normal curve   185
   4. Some applications   188
       Chapter test   191

VIII.   RANDOM VARIABLES
   1. What is a random variable   192
   2. Probability distribution of a random variable   198
   3. More examples of a random variable   201
   4. The standard deviation of a random variable   206
   5. Chebyshv's theorem   208
       Chapter test   212

IX. INFERENTIAL STATISTICS

1. Population and sample   215
2. Public opinion polls   220
3. Sampling distribution   228
4. Probability estimates   234
5. Estimating population size   237
6. Test of hypotheses   239
   Chapter test   243

X. FUN PROBLEMS

XI. MISCELLANEOUS TOPICS

1. Monte Carlo Method   253
2. Pascal's triangle   256
3. Historical Notes   259
4. Biographical Notes   264

ANSWERS   266

INDEX   304

Dependence Chart

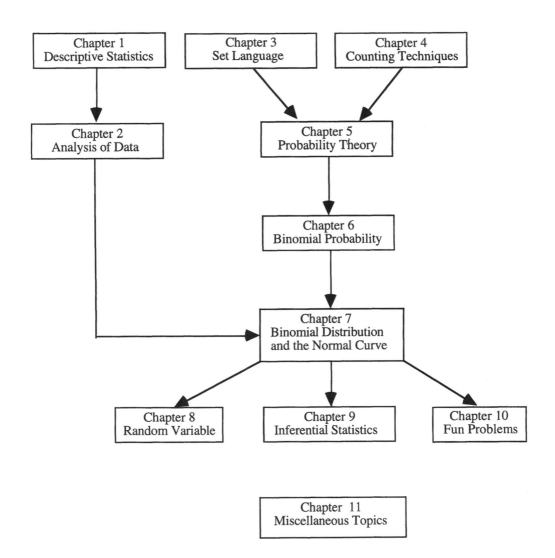

PREFACE

In this edition we have included a discussion on ranked data in the first chapter and we have streamlined chapters 3 and 4. This will help to get to probability theory fairly early. In addition to correcting some mistakes in the first edition, we have done minor rearrangements in the sections and the chapters. Other than these, the spirit and content of the book remains the same and we restate what we said in the first edition. We have produced this text keeping the following goals in mind.

(1) To make the students independent learners.

(2) To design the text for a participatory style of teaching that allows the students to actively participate in the learning process.

(3) To impress upon the students that what they learn in Finite math should be an activity for life.

(4) To make the book reasonable to the students pricewise.

Compared to other ambitious texts on Finite Mathematics that are available on the market, the present text may be described as a text for liberal arts students. Before we elaborate on how we are trying to achieve the stated goals, we must clarify what Finite Mathematics is.

Finite Mathematics includes those topics in contemporary mathematics that do not involve infinite sets, limiting processes, continuity, etc. To the best of the author's knowledge, the subject was first introduced by John G. Kemeny, J. Laurie Snell of Dartmouth college, and Gerald L. Thompson of Carnegie Institute of Technology, in a brilliantly organized book, *Introduction to Finite Mathematics,* Prentice Hall, Inc., 1956. The book was written for use at Dartmouth in a newly created course "designed to introduce a student to some concepts in modern mathematics early in his college career." The stated aim of the book was to choose topics "which are initially close to the student's experience and which have interesting and important applications to the biological and social sciences." Several topics were developed from this point of view.

The book and the kind of course created at Dartmouth became immediately popular all over America. Today, there is hardly any university or college in the country that does not offer a course on Finite Mathematics. Scores of books have appeared on the subject, all very much within the framework set by Kemeny, Snell, and Thompson. However, the subject matter has expanded considerably to include more and more topics needed by students of business, biology, the behavioral sciences, etc. In addition, the mathematical backgrounds of the students for whom such a course is designed have changed considerably. Further, different universities emphasize different topics. To suit the needs of all, books on Finite Mathematics have increased in size and price.   As a consequence, it is difficult to select a book that is suitable for a one-semester course designed primarily for liberal arts students.

Here at the University of Rhode Island, we wish to present, in a one-semester course, an introduction to the concepts and processes of modern mathematics that play an important role in the social and physical sciences of today. We try to help the students learn in a concrete way the mathematical way of thinking, the mathematical way of gathering and organizing knowledge, and the mathematical way of applying knowledge to real world problems. In many ways the quantitative and analytical thinking skills one learns in Finite Mathematics should become an activity for life. This course usually includes topics from set language, probability theory, statistics and decision theory. Such a course is of special interest to students of psychology, business, biological sciences, and to any student who wishes to have an ability to understand and use concepts and processes of modern mathematics.

To achieve our goals, we start with the basic techniques of organizing and displaying data in chapter 1. This allows the students to start the course with material that is easy to understand and that can be used almost daily while reading the newspaper, a text or while watching a baseball or basketball game. Chapter 2 is an elementary introduction to analysis of data. It is important to do these chapters in the beginning of the course so that the students may become more active in class. For example, after each test in the course, every student can himself or herself figure out how he or she is doing in the course and how the class is doing by finding the mean, mean deviation, standard deviation, etc. We have also included many examples of current interest from newspapers that were supplied by students and that concern important information on population boom, smoking, alcoholism, etc.

In chapter 3, we discuss set language and have made intrusions into domains that may not be purely finite in nature. In any case, we do find that the use of some innocent infinite sets does in fact help to achieve the goals of this course. This chapter has the dual purpose of making the student aware of the characteristic properties of different kinds of numbers. In chapter 4, we introduce the basic counting techniques. A creative teacher can use these techniques to create class activities. In the counting techniques, we are concentrating more on the kind of problems that appear in probability theory. Elementary probability theory is discussed in Chapter 5. Chapters 6 and 7 are about Binomial probability and Binomial distribution. Chapter 8 contains a simple introduction to random variables, a good knowledge of which is very important to understand the various applications of probability theory and statistics.

In probability theory, we excluded such topics as Bayes' probability, Markov chains, because we find it vain to teach these topics in a short time to the kind of students that are usually in such a course. Instead, we find it rewarding for the students and for the teacher to spend time on statistics as done in Chapters 8 and 9. In chapter 9, we have included a detailed section on public opinion polls.

Chapter 10 should help the students become independent learners. It contains some fun problems that may be solved using what they learned in the course. The problems have been arranged more or less in the order of difficulties so that a teacher may assign them to students according to their abilities. One or two of these problems may be assigned to each student to be presented in class for extra credit. The problems in this chapter are not only of historical importance but are also sources of enjoyment for the ambitious students.

Chapter 11 contains an overview of Monte Carlo techniques and some other topics. Our experiences show that the introductory discussion of random devices and Monte Carlo techniques allows the students to participate in what is going on and prepares the students for what is to come. The second part of this chapter contains the Binomial theorem via Pascal's triangle, historical notes and biographical notes. We have not included this chapter in the dependence chart because these topics may be presented whenever the teacher feels like. For example, the Monte Carlo Method does not even use the counting techinques. The later sections of this chapter may even be assigned as reading exercises.

In scope and spirit we closely follow the initiative of Kemeny, Snell, and Thompson of drawing materials initially close to the student's experience. The subject matter is presented in a straightforward manner so that the students can appreciate the beauty of mathematics. Applications are indicated wherever possible. We have made a sincere effort to set the problems and the examples on grounds that the students are likely to be familiar with.

D.K.D.

Kingston, R.I.

January, 2001

ACKNOWLEDGMENT

I would like to take this opportunity to thank all my students in my Finite Math classes who always provided the raw materials for this book. Their efforts to learn the material of the course, their ups and downs and their experiences have somehow or other filtered into this text. I am especially grateful to Jeannette Woodall, a student of Finite Math in Spring'94, for the beautiful cover design. I'd like to thank Julie D'Souza for her painstaking assistance in the reading of proofs and for her constructive efforts in improving the text.

D.K.D.

Kingston R.I.

January, 2001

SUGGESTED READINGS

Gallup, George H., *The Sophisticated Poll Watcher's Guide* , Princeton Opinion Press (1972)

Kemeny, J.G., Snell, J., Thompson, G.L., *Introduction to Finite Mathematics,* Prentice-Hall, Inc., 1974, Third edition.

Kramer, E.E., *The Main Stream of Mathematics*, Fawcett Publications, Inc., Greenwich, Conn.,1954.

Mosteller, F., *Fifty Challenging problems in Probability with solutions*, Dover publications, 1965.

Yaglom, A.M., and Yaglom, Y.M., *Challenging Mathematical Problems with Elementary Solutions* (Translated by James McCawley - revised and edited by Basil Gordon), Volume 1, Holden-Day, Inc, 1964.

National Council of Teachers of Mathematics, *Teaching Statistics and Probability*, NCTM, 1981

SUGGESTED LEARNING AIDS

A calculator, a deck of cards, three dice, four pennies, four dimes, and a quarter.

Chapter 1

ORGANIZATION AND DISPLAY OF DATA

Statistical studies based on data have become part of everyday life. News media like television, radio, newspapers and magazines constantly bombard us with so much data that some ability to understand and analyze data is needed to function in modern society. One can acquire this ability by learning the basics of descriptive statistics, namely the mathematical methods of collecting, organizing and analyzing data. There are three important things to learn:

First, one must learn to collect and work with data.

Second, one must learn to organize and display data in ways that enhance one's ability to extract information from data.

Third, one must learn to make conclusions and decisions on the basis of information gained from data.

We shall now study all these aspects of statistical methods.

1. FREQUENCY DISTRIBUTIONS

We will now discuss how to collect, organize, and display data. We first collect raw data from field work, reliable sources, observations, or by any ingenious method that one may come up with. Let's say that all of you would like to know about your classmates -- their math abilities, their verbal abilities, their likes and dislikes, their hopes and fears, and their drinking habits. We can collect the raw data by preparing a questionnaire (Figure 1.1). Ask everybody in class to complete the questionnaire and return them to the teacher.

CLASS STATISTICS

Data Collection Sheet

1.   Sex: M   F

2.   Height in inches

3.   SAT verbal score

4.   SAT math score

5.   On a scale of 0 - 10, indicate your liking for math (a score of 0 means you do not like math and a score of 10 indicates a very high degree of liking for math).

6.   On a scale of 0 - 10, indicate your fear for math (a score of 0 means you are not afraid of math and a score of 10 indicates a very high degree of fear for math).

7.   Give your best test score on a math test in high school [If the maximum possible score in the test was not 100, please give the percentage].

8.   Number of hours a week you are watch TV.

9.   Give the name of the state you come from.

10. Number of hours a week you exercise [including time you spend walking].

11. Indicate your drinking habit.

     ☐ heavy   ☐ moderate   ☐ light   ☐ rarely or never

12. Do you think you will do well in this course?

     ☐ yes   ☐ No   ☐ Don't know

Figure 1.1

A. Raw Data

Let's say you wish to study your classmates liking for math. There are several ways you can organize your study. First, collect the *raw data* from the teacher by writing down the scores from the answers to question 5 as in Figure 1.2.

5	2	1	5	6	2	4	0	0	3	5
4	7	1	3	4	0	5	2	0	5	2
3	4	0	8	5	6	6	6	4	7	

Figure 1.2

B. Frequency Distribution Table

The raw data in Figure 1.2 does not tell you anything until it is organized properly. To obtain more information from the data, you need to organize the scores in order beginning with the lowest. Then make a *frequency distribution table* using tally marks to record the number of times of each score occurs (frequency).

Score	Tally	Frequency
0	﬚	5
1	//	2
2	////	4
3	///	3
4	﬚	5
5	﬚ /	6
6	////	4
7	//	2
8	/	1

Figure 1.3. Frequency Distribution Table

DEFINITION. The *mode* of a set of data is the number that has the greatest frequency exceeding the frequency of 1.

For example, the mode of the above data is a score of 5 which appears six times.

C. Line Chart

We put each score and its frequency as an ordered pair, where the score is the first entry and the second entry is the frequency. Each of these ordered pairs is then plotted on a coordinate plane, where the horizontal axis represents test score and the vertical axis represents frequency.

The data organized in Figure 1.3 may be displayed on a *line chart* (Figure 1.4).

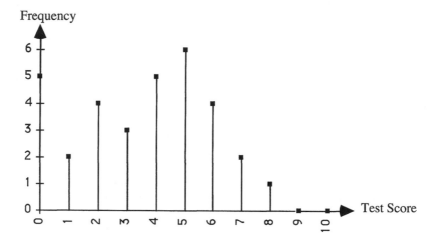

Figure 1.4

The raw data in Figure 1.2 is called *quantitative data* since each entry in the data is a number (the test scores). If a set of data refers to items that are not numbers then we call it a *qualitative data*. For example, the data in Figure 1.5 is qualitative because it refers to states. It gives the number of students from different states in a math class.

California 1   Connecticut 4   Delaware 2

New Hampshire 2   New Jersey 8   New York 9   Rhode Island 13.

Figure 1.5

D. Bar Graph

Visual representations, such as graphs, are more appealing than just a plain list of data. A popular technique of displaying data uses bar graphs. To draw the bar graph of a data, we write each item on the horizontal axis and indicate the frequencies by bars parallel to the vertical axis. It is customary to make the bars of the same width. Often people use different shades or colors to attract attention as well as to make the graph informative. Figure 1.6 is a bar graph for the qualitative data in Figure 1.5.

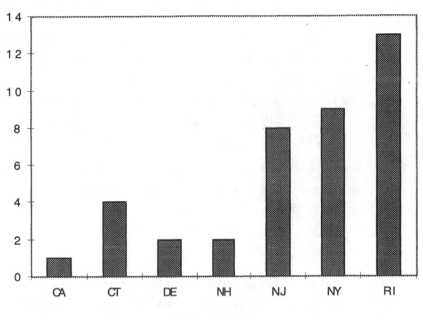

Figure 1.6

Exercises   1.1

1. (a) Display the heights of the male students in this class on a line chart.
    (b) Display the heights of the female students in this class on a line chart.

2. Display the fear for math of the students in the class on a line chart. [Collect the raw data from the teacher by writing down the scores from the responses to question 6 of the questionnaire in Figure 1.1].

3. The data below gives the scores of 24 students on a test. Organize the data into a frequency distribution table and display it on a line chart.

51	70	79	75	75	45	51	63
70	85	63	79	75	45	90	70
85	70	85	82	63	85	24	90

4. Organize the data below into a frequency distribution table and display it on a line chart.

2	5	7	0	0	1	6	8	4	3
0	5	1	4	3	9	10	2	6	1
3	2	4	4	5	7	9	9	8	0

5. A student who lives off campus decided to compare his expenses with the expenses of those who live on campus. He collected the following data about the weekly expenses in dollars from 20 students who live on campus. Organize the data into a frequency distribution table and display it on a line chart.

300	310	375	260	275	287	340	400	305	280
270	300	325	315	305	320	290	265	320	300

6. The following data represents the amount in dollars that a family of two spends on food per week for twenty weeks. Organize the data on a frequency table and draw a line chart.

40	43	65	75	70	60	48	60	40	60
45	55	50	45	50	40	45	50	45	40

7. Of the 50 students in a math class, 12 came from Michigan, 8 came from Rhode Island, 10 from Massachusetts, 12 from Connecticut, 6 from New York, and 2 from New Jersey. Display this data on a bar graph.

8. The following data gives the number of students from different states in a math class. Display this data on a bar graph.

    California 2, Connecticut 6, Delaware 2, New Jersey 8, New York 4, North Dakota 10.

9. In a 27 game basketball season, the following are the points scored by the leading scorer of a team. Organize the data on a frequency table and display it on a bar graph.

17	11	10	20	22	19	20	20	22
18	19	11	22	19	11	17	19	15
22	15	13	13	17	16	22	10	11

10. Following are the homework scores of a student. Display the data on a bar graph.

    7, 7, 8, 9, 9, 9, 10, 9, 9, 8, 7, 7, 8, 4, 5, 6, 7, 8, 8, 9

11.   Following is the grade distribution for a class last semester. Display this data on a bar graph.

4 A's, 6 B's, 14 C's, 10 D's and 4 F's

12.   Following is the grade distribution for a class last semester. Display this data on a bar graph.

3 A's, 6 B's, 12 C's, 4 D's and 9 F's

13.   Collect the raw data showing the number of students from different states in your class. Display the data on a bar graph.

2. RELATIVE FREQUENCY

An important measure that can be derived from a frequency table is the relative frequency of an item. The relative frequency of a score or an item in a set of data is obtained by dividing the frequency of the score (or item) by the total frequency.

Relative frequency and relevant statistics are very useful for comparisons and decision-making. Relative frequencies are widely used in the world of sports since they can assess the performance of a player and compare the performance with the performances of other players, past and present. For example, in baseball the statistics of players and the relative frequencies of their performances are contained on the back of baseball cards. These are not only records worth saving, but are also useful in contract decisions which usually involve millions of dollars.

Note 1.   A relative frequency is is usually less than 1 and is never greater than 1, thus a fraction. Therefore, to compare two relative frequencies, one must know how to determine which of the two fractions is larger. This can be done by either converting each fraction to decimal notation or by expressing each fraction with a common denominator. Thus,

$\frac{3}{4}$ is larger than $\frac{7}{11}$ because $\frac{3}{4} = .75$ and $\frac{7}{11} = .6363..$

or because $\frac{3}{4} = \frac{33}{44}$, $\frac{7}{11} = \frac{28}{44}$ and $33 > 28$.

Note 2. The relative frequencies of various items in data may also be expressed in percent form. The formula to use is:

$$\text{Percentage} = \text{Relative Frequency} \times 100 \qquad (1)$$

[In other words to obtain the percentage, divide the part by the whole and move the decimal point two places to the right.]

Note 3. If we know the relative frequency of an item and the total frequency then we can figure out the frequency of the item. The formula to use is:

$$\text{Frequency} = \text{Relative Frequency} \times \text{Total Frequency} \qquad (2)$$

Examples 1.1

1. The following are the scores of the students in a data. Give the relative frequency of each score.

50	50	50	53	53	53	53	53	53
60	60	60	63	63	63	63	63	63
63	63	71	71	71	71	71	71	71
80	80	80	80	89	92	92	92	

Solution. The data may be organized into the following table:

Score	Frequency	Relative Frequency
50	3	3/35
53	6	6/35
60	3	3/35
63	8	8/35
71	7	7/35
80	4	4/35
89	1	1/35
92	3	3/35

2. Organize the data from 1 above into grades: A's (90 - 100), B's (80 - 89), C's (70 - 79), D's (60 - 69), F (below 60) and give the relative frequency of each grade.

Solution.   The data may be organized into the following table:

Grade	Frequency	Relative Frequency
A	3	3/35
B	5	5/35
C	7	7/35
D	11	11/35
F	9	9/35

3. Find the percentage of each grade in the data of the previous example.

Grade	Frequency	Relative Frequency	Percentage
A	3	3/35	$\frac{3}{35} \times 100 = 8.57$
B	5	5/35	$\frac{5}{35} \times 100 = 14.29$
C	7	7/35	$\frac{7}{35} \times 100 = 20$
D	11	11/35	$\frac{11}{35} \times 100 = 31.43$
F	9	9/35	$\frac{9}{35} \times 100 = 25.71$

4.   The relative frequencies of deaths in the three most deadly occupations in America are as follows: construction workers .011, loggers .133, small plane pilots .103. Determine which is the most deadly occupation in America and find the number of deaths per 10,000 employees in the job.

Solution. Loggers is the most dangerous occupation in America because .133 is larger than .011 and .103. The frequency of deaths per 10,000 loggers is:

$$.133 \times 10,000 = 1330 \quad \text{(using formula (2))}.$$

Exercises 1.2

1.   To make the beginning students in a Finite Math Course aware of how students in the past fared in the course, the teacher read to his class the final scores of the students in the previous semester. The scores (out of 100) were:

95	88	81	95	77	85	68	53	88	77
76	66	94	72	72	53	68	73	97	72

(a) Group the data showing the grade distribution: A's (90 – 100),

     B's (80 – 89),   C's (70 – 79),   D's (60 – 69),   F's (below 60).

(b) Give the relative frequency of each grade.

(c) Display the percentage of each grade.

2. Do problem 1 if the scores were:

71	68	56	87	88	92	71	56	71	73
50	48	87	88	71	80	93	42	68	65

3. A student who does not like math has to take either Finite Math or Topics in Math. He wishes to take the course that has a better grade distribution. From the professors who taught those classes he gets the following grade distributions for the previous semester.

     Finite Math: 3 A's, 5 B's, 12 C's, 5 D's and 5 F's

     Topics in Math: 5 A's, 4 B's, 7 C's, 5 D's and 4 F's

(a) In which course the relative frequency of passing is higher?

(b) Which course should the student select if he wishes to get a C or better?

(c) Which course should the student select if he wishes to get a B or better?

(d) Which course should the student select if he wishes to get a passing grade and he knows that he could never get an A?

4. The following gives the number of home runs hit by player B in the seasons 1991 – 1995.

Year	At Bats	Home Run	Relative Frequency
1991	497	24	
1992	489	21	
1993	463	18	
1994	412	22	
1995	609	32	

(a) Determine the relative frequency of home runs for each of the given years.

(b) Judging from the relative frequency of home runs he hit, which was player B's best year and which was his worst?

5. The following table gives the number of students in this course who claim to have had math anxiety on the first day of class during the last five semesters. Give the relative frequency of students with math anxiety for each year. State if the cases of math anxiety are increasing or decreasing.

Semester	Number of students with Math Anxiety	Total number of students	Relative frequency
Fall'93	3	32	
Spring'94	7	34	
Fall'94	15	68	
Spring'95	10	36	
Fall'95	15	53	

6.   Determine the relative frequency of

      (a) The letter e in the word *experience*.

      (b) The letter i in the word *antidisestablishment*.

7.   *(Deadly Occupations)*. The relative frequencies of deaths per hundred in the five most deadly occupations in America are as follows: construction workers .011, loggers .133, small plane pilots .103, taxi drivers .050, and trucking .026. Determine which is the least deadly among the five most deadly occupations in America and find the number of deaths per 100,000 employees in the job.

3. CIRCLE GRAPHS OR PIE CHARTS

Data may also be displayed on a circular region. The idea is to divide a circular region into convenient sectors and to assign to each item in the data a sector proportionate to the data. This is quite easily done if one knows how to use a protractor, remembers that one half of the circular region corresponds to $180^{\circ\circ}$ and remembers that the whole circle corresponds to 360° (Figure 1.7). The result is a circle graph or a pie chart. Pie charts are convenient when the relative size of items in data are to be emphasized.

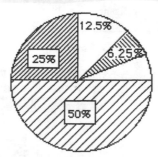

Figure 1.7

Example 1.2.   Students' Finances

Sources: *Where it comes from*

parents	$7,000.00
financial aid	$4,000.00
county scholarship	$1,500.00
summer job	$4,500.00
gifts from relatives	$1,000.00
Total:	$18,000.00

Expenses: *Where it goes*

tuition	$4,000.00
room & board	$5,000.00
books, clothes, and stationary	$1,000.00
entertainment (including travel)	$3,500.00
booze and smokes	$1,000.00
wheels (including parking and speeding tickets)	$3,500.00
Total:	$18,000.00

Figure 1.8

The data in Figure 1.8 may be displayed on circular regions as done in Figure 1.9.

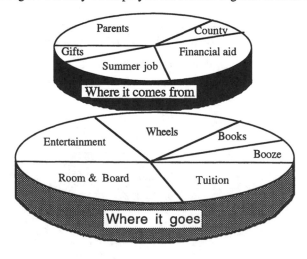

Figure 1.9

The charts in Figure 1.9 are known as *circle graphs* or *pie charts*. In each of these, $18,000.00 corresponds to the entire circle, which has 360°. To find the dollar amount assigned to one degree, take the dollar amount for the entire circle and divide it by 360, the number of degrees at in a center. So,

$$1° \text{ corresponds to } \frac{18000}{360} = \$50.00.$$

Therefore, 10° corresponds to $500.00, 70° corresponds to $3,500.00, and so on. The following formula is useful in making pie charts.

Formulas for the Size of a Pie

Since 360° corresponds to 100%, we have

$$1° \text{ corresponds to } \frac{100}{360} = .28\% \text{ (approximately) and}$$

$$1\% \text{ corresponds to } \frac{360}{100} = 3.6°.$$

Examples 1.3

1.   On a pie chart find the percentage that corresponds to a 5°.

Solution. $\frac{100}{360} \times 5 = 1.39\%$.

2.   Find the angle at the center of a pie chart corresponding to 10%

Solution. $\frac{360}{100} \times 10 = 36°$.

3.   Of the 32 students in a class 12 came from Rhode Island, 4 from New York, 8 from Massachusetts, 6 from Connecticut and 2 from New Jersey. Find the percentages of students from each of the states and display them on a pie chart.

Solution. Percentage of students from RI $= \frac{12}{32} = .375 = 37.5\%$.

Percentage of students from MA = 25%
Percentage of students from CT = 18.75%
Percentage of students from NY = 12.5%
Percentage of students from NJ = 6.25%

Figure 1.10

Exercises 1.3

1. A student spends 50% of his time sleeping and taking care of himself, 25% socializing, 12.5% in recreational activities, and the rest in his studies. Display the data on a pie chart and give the percentage of time he spends studying.

2. A student spends 37.5% of his time sleeping and taking care of himself, 12.5% socializing, 6.25% in recreational activities, and the rest in his studies. Display the data on a pie chart and give the percentage of time he spends studying.

3. In 1987, states paid roughly 50% of the $211.6 billion in the year's public education expenses, local governments paid 43.75% of the education tab and federal government covered the remaining 6.25%. Display the data on a pie chart.

4. 50% of the total expenses of a student go to room and board, 18.75% on books and stationary, 6.25% on entertainment, and the rest on miscellaneous items. Display the data on a pie chart and give the percentage of amount spent on miscellaneous items.

5. A student receives 37.5% of his money from his parents, 25% from summer jobs, 12.5% from scholarships, 6.25% from gifts, and the rest from work-study programs. Display the data on a pie chart and give the percentage of amount that comes from work-study programs.

6. According to the National Center for Health Statistics, the following is the percentage of the male population of smokers, nonsmokers, and former smokers for the years 1977-1985. For each year listed, create a pie chart that displays the data for that year.

Year	Smokers	Nonsmokers	Former smokers
1977	41%	31%	28%
1980	38%	33%	29%
1983	36%	34%	30%
1985	33%	36%	31%

7. According to the National Center for Health Statistics, the following is the percentage of the female population of smokers, nonsmokers, and former

smokers for the years 1977-1985. For each year listed, create a pie chart that displays the data for that year.

Year	Smokers	Nonsmokers	Former smokers
1977	32%	54%	14%
1980	29%	55%	16%
1983	29%	55%	16%
1985	28%	54%	18%

8.   Out of every dollar a city's budget receives, 50 cents come from general sources, 25 cents from federal aid, 12.5 cents from state aid, 6.25 cents from property tax, and the rest from capital. Out of every dollar the city spends, 37.5 cents are spent on personnel, 25 cents on education, 12.5 cents on administration of justice, 12.5 cents on health, 6.25 cents on fire, and the rest on miscellaneous expenses. Draw two pie charts, one displaying the city's budget dollar - where it comes from, and the other displaying where the city's budget dollar goes. (See Figure 1.9).

9.   On a pie chart find the percentage that corresponds to each of the following angles:

   (a) 10°   (b) 180°   (c) 20°   (d) 45°

10.   Find the angle at the center of a pie chart corresponding to each of the following:

   (a) 5%   (b) 20%   (c) 12.5%   (d) 45%

11.   Of the 40 students in a math class, 12 came from RI, 10 from MA, 8 from CT, 6 from NY, and 4 from NJ. Find the percentages of students from each of the five states and display them on a pie chart.

12.   Of the 50 students in a math class, 15 came from VA, 12 from SC, 10 from MI, 8 from IL, and 5 from PA. Find the percentages of students from each of the five states and display them on a pie chart.

4. STEM AND LEAF DISPLAYS

Frequency distribution tables and line charts are rather elaborate methods of organizing and displaying data. Often the preparations are very time consuming. Therefore, to take a quick look at the nature of data, a useful method is to organize the data on a stem and leaf display. To do so, take the left most digit of the data as the stem and the other digits as the leaves. The stem and the leaves are separated by empty spaces or a vertical line. Once a stem and leaf is constructed, we can determine various characteristics of the data. We illustrate this with some examples.

Examples 1.4

1.   *(Symmetric data)* The following data is from the responses to question 7 of the questionnaire by a Finite Math class.

45, 59, 13, 21, 46, 42, 48, 47, 64, 22, 75, 34, 35, 51, 53, 46, 63, 32.

Solution.

1	3
2	1 2
3	4 5 2
4	5 6 2 8 7 6
5	9 1 3
6	4 3
7	5

The above data is symmetric about the middle (the 40's) because it has as many scores above as below the middle leaf and the scores are distributed symmetrically. The mode of the data is 46.

2.   *(Skewed data.)* Display the following data on a stem and leaf and discuss some characteristics of the data.

67	55	71	60	73	71	57	44	67	98
67	77	90	71	63	92	71	57	53	53
96	80	85	89	67	67	71	80	85	57
50	67	50	60	63					

Solution. The stems are the digits in the tens 4, 5, 6, 7, 8 and 9 because the lowest score is 44 and the highest score is 98. The leaves are the digits in the units. The display will look like:

4	4
5	5 7 7 3 3 7 0 0
6	7 0 7 7 3 7 7 7 0 3
7	1 3 1 7 1 1 1
8	0 5 9 0 5
9	8 0 2 6

From the stem and leaf display, we can see that the mode of the given data is 67, the scores are concentrated in the 50's, 60's and 70's, and there are quite a few high scores and not many low scores. The data is skewed towards the high values.

3.　　*(Bimodal data)* Construct a stem and leaf display of the following data and discuss some of its characteristics.

　　　57　43　45　92　89　76　61　52　51　34　32　74　85
　　　80　96　84　40　97　60　74　75　41　58　81　46　87.

Solution.

3	4 2
4	3 5 0 1 6
5	7 2 1 8
6	1 0
7	6 4 4 5
8	9 5 0 4 1 7
9	2 6 7

The data shows that the students fall mainly into two groups: one group that is doing very well and the other that is not doing well. There are very few who may be called average. Such a data that contains two distinct groups is called *bimodal*. If the scores in a test of a class show this kind of data, then the teacher has the difficult task of teaching to two groups of varying abilities.

4.　　The following data gives the SAT verbal scores of the students in a finite math class. Construct a stem and leaf display of the data and discuss some of its characteristics.

700, 655, 465, 260, 330, 270, 450, 630, 640, 555, 520, 710, 420, 430, 700, 560, 345, 550, 530, 375

Solution. The display has hundreds' digits as the stems and the leaves are the numbers formed by the digits in tens and units.

2	60 70
3	30 45 75
4	65 50 20 30
5	55 20 60 50 30
6	55 30 40
7	00 10 00

5.   *(Discarding digits that are not of importance.)* Another advantage of a stem and leaf display is that we may discard digits that do not really affect the display. For example, we may discard the units in the preceding example and display it as follows:

2	6 7
3	3 4 7
4	6 5 2 3
5	5 2 6 5
6	5 3 4
7	0 1 0

Here the hundreds' digits are the stems and the tens' digits are the leaves. The above data is almost symmetric and the mode is 550.

In general, if the stem and leaf display of the data has as many scores above as below the middle leaf and they are distributed symmetrically about the middle then the data is symmetric. The data is *skewed* if numbers are concentrated on the high or low end of the stem. The data is *bimodal* if the leaves show two distinct groups. These characteristics are easy to observe from stem and leaf displays. In the next section, we will learn other methods to describe these characteristics.

6.   *(Subdividing stems)* If one or more leaves are too long or if there are only a few stems, we may subdivide the leaves as illustrated in the following example.

5	2
6	7 0 6 7 3 1 7 4 0
7	1 3 1 7 2 1 9 5 7 7 4 3 6
8	0 5 9 0 4 8 0 2 7

The above diagram may be displayed by subdividing the stems by using two leaves for each stem (one for 0 - 4 and the other for 5 - 9) as follows:

```
5  |  2
5  |
6  |  0  3  1  4  0
6  |  7  6  7  7
7  |  1  3  1  2  1  4  3
7  |  7  9  5  7  7  6
8  |  0  0  4  0  2
8  |  5  9  8  7
```

Exercises 1.4

1. The following are the test scores of a math class. Construct a stem and leaf display of the data and determine if it is symmetric, bimodal or skewed.

    78, 82, 89, 82, 35, 52, 63, 76, 88, 93, 97, 47, 62, 77, 84,
    39, 97, 88, 87, 63, 70, 63, 64, 77, 21, 75, 86, 81, 51, 67.

2. The following are the test scores of a math class. Construct a stem and leaf display of the data and determine if it is symmetric, bimodal or skewed.

    68, 92, 89, 72, 55, 32, 33, 76, 98, 93, 77, 87, 42, 47, 84, 39,
    97, 89, 88, 63, 60, 73, 64, 87, 51, 75, 96, 84, 61, 57, 47, 45.

3. The following are the SAT verbal scores of the students in a finite math class. Construct a stem and leaf display of the data and determine if it is symmetric, bimodal or skewed.

    540, 230, 240, 320, 500, 610, 290, 310, 300, 270, 410, 340, 250, 260, 340,
    470, 420, 420, 330, 340, 390, 395, 295, 360, 365, 375, 400.

4. The following are the SAT math scores of the students in a finite math class. Construct a stem and leaf display of the data and determine if it is symmetric, bimodal or skewed.

    300, 330, 230, 480, 400, 310, 290, 310, 250, 270, 400, 340, 350, 260, 320,
    370, 420, 400, 330, 340, 300, 295, 295, 360, 350, 355, 300, 450, 470.

5. Construct a stem and leaf display of the best math test scores of your classmates [from the responses to question 7 of the questionnaire].

6. Collect the SAT verbal scores of your classmates [from the responses to question 3 of the questionnaire]. Construct a stem and leaf display of the data and determine if it is symmetric, bimodal or skewed.

7. Collect the SAT math scores of your classmates [from the responses to question 4 of the questionnaire]. Construct a stem and leaf display of the data and determine if it is symmetric, bimodal or skewed.

5. GROUPED DATA

It is possible to shorten the frequency distribution by grouping the data into convenient intervals. This results in a *frequency distribution with grouped data*. Grouped data may not be as detailed as the frequency distribution, but may provide additional information. We illustrate this with the following raw data given in 2 of Examples 1.4.

67	55	71	60	73	71	57	44	67	98
67	77	90	71	63	92	71	57	53	53
96	80	85	89	67	67	71	80	85	57
50	67	50	60	63					

Instead of finding the frequency of each score, select convenient intervals, for example 40 - 45, 46 - 50, . . ., 96 -100. The first number of the first interval and the last number of every interval give the points of partition on the number line [Figure 1.11].

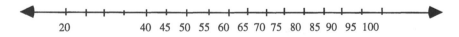

20   40 45 50 55 60 65 70 75 80 85 90 95 100

Figure 1.11

The *length* of each interval is the difference between the two points of partition that contain the interval. The length of the interval 46 - 50 is 5 since the two points of partitions containing this interval is 45 and 50. Tally the frequency of each score to get the following:

Interval	Tally	Frequency
40 - 45	/	1
46 - 50	//	2
51 - 55	///	3
56 - 60	////	4
61 - 65	///	3
66 - 70	~~////~~ /	6
71 - 75	~~////~~ /	6
76 - 80	///	3
81 - 85	//	2
86 - 90	//	2
91 - 95	/	1
96 - 100	//	2

Figure 1.12. Frequency Distribution With Grouped Data

Histogram

If the bars of a bar graph of a frequency distribution with grouped data are drawn without any gap between two consecutive bars then the graph is called a *histogram*.

To draw a histogram, take the horizontal axis to represent test scores and the vertical axis to represent frequencies. Then, the width of a bar represents the lengths of the selected intervals (these intervals are usually of the same length).

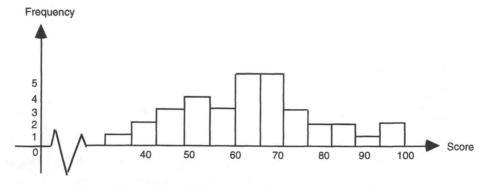

Figure 1.13. Histogram

Frequency Polygon

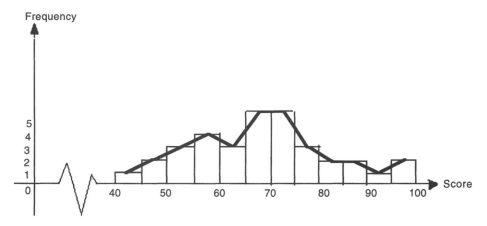

Figure 1.14. Frequency Polygon

The *frequency polygon* for a set of data is constructed by joining consecutive midpoints of the tops of the bars of the histogram with line segments the [Figure 1.14.].

Exercises 1.5

1. Draw a histogram of the heights of students in this class (consider intervals of four inches).

2. Divide the interval 0 - 10 into five equal intervals. Group the scores (of liking math) of this class into these intervals. Draw a histogram and a frequency polygon, and based on the data, tell your teacher whether he would have

    (a) fun,   (b) a hard time, or   (c) a very hard time teaching this class.

3. The data below gives the homework scores of students. Construct a frequency distribution table with grouped data by considering intervals of length 2. Display the grouped data on a histogram.

2	5	7	0	0	1	6	8	4	3
0	5	1	4	3	9	10	2	6	1
3	2	4	4	5	7	9	9	8	0

4. For the following data, make a frequency distribution table with grouped data by taking intervals of length 20. Display the data on a histogram.

300	310	375	360	275	287	340	400	305	280
270	300	325	315	365	380	290	265	320	300

5.   The data below gives the scores of 24 students on a test. Construct a frequency distribution table with grouped data by considering grades:

F (below 60), D's (60 - 69), C's (70 - 79), B's (80 - 89), A's (90 - 100).

Display the grouped data on a histogram.

51	74	79	75	73	45	58	63
70	86	62	79	73	47	90	70
87	72	85	82	61	86	24	90

6.   The data below gives the scores of students on a test. Construct a frequency distribution table with grouped data by considering grades:

F (below 60), D's (60 - 69), C's (70 - 79), B's (80 - 89), A's (90 - 100).

Display the grouped data on a histogram.

41	72	78	75	73	55	68	63
70	86	62	79	63	67	90	70
85	73	86	84	61	86	94	90

7.   The data below gives the SAT Math scores of students in a Finite math class. Construct a frequency distribution table with grouped data by considering intervals of 50. Display the grouped data on a histogram.

550	640	570	480	590	510	450	300	310	510
580	420	570	470	330	470	540	390	340	360
470	620	440	465	520	380	460	500	520	570

8.   The data below gives the SAT Verbal scores of students in a Finite math class. Construct a frequency distribution table with grouped data by considering intervals of 50. Display the grouped data on a histogram.

320	470	340	400	300	460	350	510	370	630
500	420	560	440	450	610	510	650	500	510
680	560	310	390	410	520	320	700	390	400

9.   The following data represents the amount in dollars that a family of two spends on food per week for twenty weeks. Make a frequency distribution table with grouped data by considering intervals of length 5. Display the data on a histogram.

| 40 | 43 | 65 | 75 | 70 | 60 | 48 | 60 | 40 | 60 |
| 45 | 55 | 50 | 45 | 50 | 40 | 45 | 50 | 45 | 40 |

10.   In a 27 game basketball season, the following are the points scored by the leading scorer of a team. Make a frequency distribution table with grouped data by considering intervals of length 5. Display the data on a histogram.

17	11	10	28	23	19	20	26	22
18	29	36	44	29	31	27	29	25
22	15	23	23	27	26	22	10	31

11.   The following data are the daily low temperatures in Fahrenheit for the month of February beginning with February 1. Construct a frequency distribution table with grouped data by considering intervals of length 5. Display the data on a histogram.

9	2	10	12	15	28	16	12	13
5	29	20	26	17	8	22	10	25
19	39	40	36	24	28	23	19	8

Cumulative Frequency

Suppose you are in a big class where there are 300 students. You need to study the test scores of the class, and you wish to ask the teacher to scale the grades. To present your case, you need to show the teacher how many students are flunking (getting less than 60%), how many are getting C's (less than 70%), etc. Also, you wish to convince the teacher to lower the cut-off marks for grades. For example, you may say that if he lowers the passing grade to 55%, then half the class will not flunk.

Interval	Frequency	Cumulative Frequency
40 - 45	15	15
46 - 50	7	22
51 - 55	10	32
56 - 60	20	52
61 - 65	17	69
66 - 70	11	80
71 - 75	12	92
76 - 80	4	96
81-85	2	98
86-90	2	100
91-95	0	100
96-100	0	100

Figure 1.15

To be prepared with your statistics, you will need to know the *cumulative frequency* up to and including each interval. This is done by adding to the table of frequency distribution with grouped data a separate column to indicate the cumulative frequency. The entry in this column corresponding to any interval is obtained by adding the frequency of that interval to the sum of the frequencies of the intervals that preceded the given interval. Figure 1.15 organizes, with cumulative frequency, the test scores of 100 students on a math test.

Cumulative Frequency Polygon

We can make a visible representation of cumulative frequency by plotting the ordered pairs where the first entry is the end point of an interval, and the second entry is the cumulative frequency for that interval. Thus, from the given table we can plot the points (45,15), (50,22), (55,32), and so on. By joining consecutive points with line segments, we get the *cumulative frequency polygon* for the given data. The cumulative frequency polygon for the data in Figure 1.15 is in Figure 1.16.

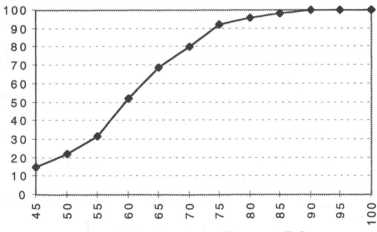

Figure 1.16. Cumulative Frequency Polygon

Exercises 1.6

For each of the data in problems 1 - 4,

        (a) Make a table indicating frequency and cumulative frequency.

        (b) Construct a frequency polygon.

        (c) Construct a cumulative frequency polygon.

1.   Intervals of length 2.

2	5	7	0	0	1	6	8	4	3
0	5	1	4	3	9	10	2	6	1
3	2	4	4	5	7	9	9	8	0

2.   Intervals of length 20.

| 300 | 310 | 375 | 260 | 275 | 287 | 340 | 400 | 305 | 280 |
| 270 | 300 | 325 | 315 | 305 | 320 | 290 | 265 | 320 | 300 |

3.   Intervals of length 50.

550	640	570	480	590	510	450	300	310	510
580	420	570	470	330	470	540	390	340	360
470	620	440	465	520	380	460	500	520	570

4.   Intervals of length 50.

`320	470	340	400	300	460	350	510	370	470
500	420	560	440	450	610	510	650	500	510
680	450	310	390	410	520	320	300	390	400

5.   The following data gives the scores of students in a math test. Group the data into intervals of length 10.

51	74	79	35	23	55	58	33	57	75
70	86	62	79	53	47	90	70	42	54
87	72	85	82	61	86	24	90	46	94.

(a)   Make a table indicating frequency and cumulative frequency.

(b)   How many students fail if the passing grade is 61 or better?

(c)   How many students fail if the passing grade is lowered to 51 or better?

6.   The following data gives the scores of students in a math test. Group the data into intervals of length 10.

42	54	69	45	63	45	58	63	85	48	90
86	52	59	53	57	88	70	37	54	77	62
55	92	61	56	34	90	53	63.			

(a)   Make a table indicating frequency and cumulative frequency.

(b)   How many students fail if the passing grade is 61 or better?

(c)   How many students fail if the passing grade is lowered to 51 or better?

6. RANKED DATA

If a given data set is ranked in increasing order, we get a ranked data. For convenience, a ranked data may be grouped in several different ways.

Median. The *median* of a ranked data set divides the data into two equal parts. A ranked data has one median. We shall refer to the elements in the data that are less than the median as the first half and the elements that are greater than the median as the second half of the data.

To find the median of a ranked data with an EVEN number of elements, divide the total number of elements in the data by 2 - the result will be a whole number. The median of the data is the average of that and the next term in the data.

To find the median of a ranked data with an ODD number of elements, divide the total number of elements in the data by 2, round the result to the next whole number. The median is that term in the data.

Examples 1.5

1.   Find the median of 2, 7, 11, 13, 14, 17, 18, 19, 21, 23.

solution. This data has 10 elements. Dividing 10 (an even number) by 2, we get 5, so the median is the average of the fifth and the sixth terms in the data, that is,

$$\text{Median} = \frac{14 + 17}{2} = 15.5.$$

2.   Find the median of 0, 3, 5, 11, 14, 16, 18, 19, 23.

solution. This data has 9 elements. Dividing 9 (an odd number) by 2, we get 4.5. The whole number next to 4.5 is 5, so the median is the fifth term in the data, that is, 14.

Quartiles. Quartiles divide a ranked data set into four equal parts. A ranked data set has three quartiles. They are called the first, the second and the third quartile, denoted by Q_1, Q_2, and Q_3, respectively. The second quartile Q_2 is the same as the median of the data. The first quartile Q_1 is the median of the data that are in the first half and Q_3 is the median of the second half.

Examples 1.6

1.   Find the median, Q_1, Q_2 and Q_3 of the following data:

$$14, 34, 26, 22, 17, 12, 13, 23, 12, 10, 4, 22, 28, 26, 2, 6, 1, 3.$$

Solution. The data may be ranked as:

$$1, 2, 3, 4, 6, 10, 12, 12, 13, 14, 17, 22, 22, 23, 26, 26, 28, 34.$$

Since there are 18 elements in the data, the median falls between the ninth and the tenth terms. So, the median is $\dfrac{13 + 14}{2} = 13.5$, which is also Q_2. The first half contains nine terms, so Q_1 is the fifth term, namely, 6. In the same way the third quartile Q_3 falls on the fifth term of the second half. So, $Q_3 = 23$.

Percentiles. Percentiles divide a ranked data set into 100 equal parts. A data set has 99 percentiles. The first percentile is denoted by P_1 and about 1% of the elements in the data are less than P_1. The second percentile is denoted by P_2 and about 2% of the elements in the data are less than P_2. In general, The kth percentile is denoted by P_k. About k% of the elements in the data are less than P_k, and the remaining are larger than P_k in value. The 50th percentile or P_{50} is the same as the median, P_{25} is the same as the first quartile and P_{75} is the same as the third quartile. Percentiles are useful for large data sets like the SAT scores.

To find the percentile P_k for any k,

Step-1. Find $\dfrac{n}{100}$, where n is the number of elements in the data.

Step-2. Multiply $\dfrac{n}{100}$ by k , that is, find $k \times \dfrac{n}{100}$. If this is a whole number then P_k falls on the mean of that and the next term in the ranked data. If $k \times \dfrac{n}{100}$ is not a whole number, round it up to the next whole number and P_k falls on that term in the ranked data.

Examples 1.7

1.   Consider a ranked data containing 200 terms. Determine the positions of each of the following percentiles.

(a) P_1 (b) P_2 (c) P_3 (d) P_{25}

(e) P_{50} (f) P_{75} (g) P_{90} (h) P_{99}

Solution. We have $\dfrac{n}{100} = \dfrac{200}{100} = 2$. So,

(a) P_1 is the mean of the 2nd and the 3rd terms,

(b) P_2 is the mean of the 4th and the 5th terms (since $2 \times 2 = 4$),

(c) P_3 is the mean of the 6th and the 7th term since $3 \times 2 = 6$.,

(d) P_{25} is the mean of the 50th and the 51st terms,

(e) P_{50} (the median) is the mean of the 100th and the 101st terms, and so on.

2.   Consider a ranked data containing 201 terms. Determine the positions of each of the following percentiles.

(a) P_1 (b) P_2 (c) P_3 (d) P_{25} (e) P_{50}

(f) P_{75} (g) P_{90} (h) P_{99}

Solution. We have $\dfrac{n}{100} = 2.01$; the next whole number is 3 so,

(a) P_1 is the third term,

(b) P_2 is the fifth term (since $2 \times 2.01 = 4.02$),

(c) P_3 is the seventh term since $3 \times 2.01 = 6.03$,

(d) P_{25} is the 51st term,

(e) P_{50} (the median) is the 101st terms, and so on.

3.   Find (a) P_5, (b) P_{10}, (c) P_{25}, (e) P_{50}, (f) P_{75}, (g) P_{90}, and (h) P_{99} for the following ranked data set.

$$38 \quad 49 \quad 50 \quad 50 \quad 55 \quad 57 \quad 61 \quad 64 \quad 67 \quad 69 \quad 70 \quad 71$$
$$71 \quad 72 \quad 74 \quad 75 \quad 75 \quad 76 \quad 80 \quad 83 \quad 88 \quad 94 \quad 98 \quad 99.$$

Solution.   Since $n = 24$, $\dfrac{n}{100} = .24$. Since this is not a whole number, the first percentile of significance is P_5. Since $5 \times .24 = 1.20$, P_5 is the second term. $P_5 = $ Fifth percentile $= 49$. In this case, 5 % of the scores are less than 49.

Similarly, to find the 10th percentile, we have $10 \times .24 = 2.4$. So,

$P_{10} = $ Tenth percentile $= 50$.

$P_{25} = $ 25th percentile $= \dfrac{57 + 61}{2}$ (since $25 \times .24 = 6$, P_{25} is the mean of the

sixth and the seventh terms)

$P_{90} = 90$ th percentile $= 94$ (since $90 \times .24 = 21.6$, P_{90} is the 22nd term).

$P_{99} = 99$th percentile $= 99$ (since $99 \times .24 = 23.76$, P_{99} is the 24 th term).

It is to be noted that some times (for example, when the data set is small) some of the percentiles may give an approximate percentage rather than an exact percentage.

Exercises 1.7

1. Find the median, Q_1, Q_2 and Q_3 of the following data.

33	21	37	42	45	49	48	30	26	24
29	35	36	27	26	44	40	19	20	25.

2. Find the median, Q_1, Q_2 and Q_3 of the following data.

56	45	98	71	78	94	88	65	59	43
69	82	87	75	68	63	64	66	91	90.

3. Find the median, Q_1, Q_2 and Q_3 of the following data:

19	23	15	14	27	35	49	48	45	33	33
44	37	29	28	36	42	31	22	34	47.	

4. Find the median, Q_1, Q_2 and Q_3 of the following data.

39	43	25	24	27	15	19	43	54	13	34
44	57	27	58	36	52	11	32	45	27.	

5. Consider a ranked data containing 400 terms. Determine the positions of each of the following percentiles.

   (a) P_1 (b) P_2 (c) P_3 (d) P_{25} (e) P_{50}
   (f) P_{75} (g) P_{90} (h) P_{99}.

6. Consider a ranked data containing 401 terms. Determine the positions of each of the following percentiles.

   (a) P_1 (b) P_2 (c) P_3 (d) P_{25} (e) P_{50}
   (f) P_{75} (g) P_{90} (h) P_{99}.

7. Your teacher tells you that in the test you scored in the ninety-fifth percentile.

   (a) Approximately what percentage of the test scores would be less than yours?
   (b) If 80 students took the test, approximately how many students' scores are less than yours?

8.   Your teacher tells you that in the test you scored in the eighty-fifth percentile.

      (a) Approximately what percentage of the test scores would be less than yours?

      (b) If 60 students took the test, approximately how many students' scores are less than yours?

9.   Find (a) P_5,   (b) P_{10},   (c) P_{25},   (d) P_{50},   (e) P_{75} for the following ranked data set.

25	42	44	50	52	55	58	61	62	63	64
65	68	72	73	74	74	75	77	78	79	80
81	82	83	84	85	85	86	87	88	89	90
91	92	93	96	97	99	99.				

10.   Find (a) P_5,   (b) P_{10},   (c) P_{25},   (d) P_{50}, and (e) P_{90} for the following ranked data set:

11	12	15	24	27	28	35	43	47	48
49	53	59	59	62	63	64	66	72	75
79	80	81	83	83	86	88	89	95	96.

*1.7 BASEBALL AND BASKETBALL

Exercises 1.8

1.   The following gives the batting record of a player A for the last season.

Year	AB	R	H	2B	3B	HR	RBI	SB	BB	SO
1988	590	86	192	39	8	22	119	16	87	38

      (a) Determine the relative frequency of each category. (Use the number of at bats as the total frequency.)

      (b) Determine his BA (*Batting average* = relative frequency of H = hits) $= \dfrac{192}{590}$.

      (c) Determine his TB (Total base) and SAVG (Relative Frequency of TB, also called *Slugging average* and often denoted by SLG).

(d) Use these relative frequencies to estimate the number of hits Mike is likely to hit in the 1989 season in 500 'at bats'.

AB = At bats	R = Runs scored	H = Hits
2B = Double	3B = Triple	HR = Home Run
RBI = Runs batted in	SB = Stolen bases	
BB= Bases on Balls	SO = Strike outs.	

In calculating TB, a HR counts as 4 bases, each 3B as 3 bases, and each 2B as 2 bases and since each of these is also counted once in calculating H, the formulas to determine TB and SAVG are:

$$TB = H - (2B + 3B + HR) + 2 \times (2B) + 3 \times (3B) + 4 \times (HR)$$

$$= H + (2B) + 2 \times (3B) + 3 \times (HR)$$

$$SAVG = TB \div AB = \frac{H + (2B) + 2 \times (3B) + 3 \times (HR)}{AB} .$$

2. The following gives the batting record of player B for the last season.

Year	AB	R	H	2B	3B	HR	RBI	SB	BB	SO
1988	540	93	159	37	5	18	92	25	62	89

(a) Determine the relative frequency of each category. (Use the number of 'at bats' as the total frequency).

(b) Determine his BA (batting average).

(c) Determine his SAVG.

(d) Of the two players, A (see Problem 6) and B, who do you think performed better during the last season?

3. The following gives the pitching record of Roger Clemens for the 1988 season.

Year	G	IP	W	L	R	ER	SO	BB
1988	35	264	18	12	93	86	291	62

where G = Games played   IP = Innings Pitched   W = Win

L = Loss   R = Runs charged   ER = Earned Run

SO = Strike Outs   BB = Bases on Balls.

(a) By using the IP as the total frequency, determine the relative frequency of each category.

(b) Determine the ERA. The ERA (earned run average) is roughly the ER per game. Since every game is usually of 9 innings, the formula to determine the ERA is: $9 \times$ (Relative frequency of ER) $= 9 \times \dfrac{ER}{IP} = 9 \times \dfrac{86}{264} = 2.93$.

4. The following gives the pitching record of Roger Clemens for the 1989 season.

Year	G	IP	W	L	R	ER	SO	BB
1989	31	228.3	21	6	59	49	209	54

(a) Determine the relative frequency of W, L, R, ER, SO and BB.

(b) Determine his ERA for the 1989 season.

(c) Determine whether his performance in the 1989 season was better than his performance in the 1988 season.

5. The following gives the number of home runs hit by player A in the last five seasons, and their relative frequencies. (To determine the relative frequency, we have used the number of 'at bats' as the total frequency.)

Year	At Bats	Home Run	Relative Frequency
1991	451	17	.038
1992	573	24	.042
1993	626	39	.062
1994	657	28	.043
1995	546	27	.049

6. The following is the number of home runs hit by player B in the years 1991 - 95.

Year	At Bats	Home Run	Relative Frequency
1991	497	24	
1992	489	21	
1993	463	18	
1994	412	22	
1995	609	32	

(a) Determine the relative frequency of home runs for each of the given years.

(b) Judging from the relative frequency of home runs he hit, which was player B's best year and which was his worst?

7.   The following is the free throw statistics of five players of a basketball team.

Player No	Free throws attempted	Free throws made
1	40	37
2	60	40
3	50	38
4	70	45
5	35	30

(a) Determine the relative frequency of successful free throws made by each of the players.

(b) Can the opposing side use the data to determine which player to foul in a tight situation when the team is already in a one-and-one situation? If so, decide which player they should foul.

8.   The following is the free throw statistics of the five players of a basketball team.

Player No	Free throws attempted	Free throws made
1	45	33
2	65	45
3	50	35
4	70	49
5	35	28

(a) Determine the relative frequency of free throws made by each of the players.

(b) Can the opposing side use the data to determine which player to foul in a tight situation when the team is already in a one-and-one situation? If so, decide which player they should foul.

CHAPTER TEST 1

1.   (a)   On a pie chart find the percentage that corresponds to a $45°$ pie.

(b)   Find the angle at the center of a pie chart corresponding to 15%.

2.   The following is the grade distribution for Finite Math and Topics in Math courses last semester.

Finite Math: 5 A's, 10 B's, 20 C's, 3 D's and 2 F's

Topics in Math: 5 A's, 6 B's, 12 C's, 4 D's and 3 F's

Organize the raw data indicating the relative frequency of the grades for each course.   In which course the relative frequency of getting an A or B is higher?

3. Of the 40 students in a math class 20 came from Rhode Island, 10 from Massachusetts, 5 from Connecticut, 3 from New York and 2 from New Jersey. Find the percentages of students from each of the five states and display them on a pie chart.

4. Find the median, Q_1, Q_2 and Q_3 of the following data:

21   23   19   23   30   20   21   22   32   41   35
19   17   34   43   27   18   26   24   38   50   35

5. Construct a stem and leaf display for the following data and find if it is symmetric, bimodal or skewed.

64   67   68   70   72   27   28   26   24   38   50   35
31   41   51   52   32   42   30   44   75   68   43   38.

6. Find (a) P_5, (b) P_{10}, (c) P_{25}, (d) P_{50}, (e) P_{75} for the following ranked data set:

1	2	2	3	4	5	5	6	7	8	9	9	9	10
10	11	12	13	14	14	15	16	17	18	18	19	20	
20	21	22	23	24	25	26	27	29	30	32	35	36	

Problems 7 - 10 refer to the following raw data that gives the scores in a test:

40	65	45	80	85	60	60	65	60	50
50	55	60	60	70	61	43	73	70	42
45	67	92	74	86	95	83	78	96	85

7. Organize the data giving the frequency distribution and draw a line chart for the top fifteen scores in the raw data.

8. Make a frequency distribution table with grouped data by considering intervals of length 5 and draw a histogram.

9. Make a table indicating the frequency and the cumulative frequency of each interval from problem 8.

10. (a) Construct a frequency polygon from the table you made for problem 8.
    (b) Construct a cumulative frequency polygon from the table you made for problem 9.

Chapter 2

ANALYSIS OF DATA

1. MEASURES OF CENTRAL TENDENCY

When analyzing data, it is of fundamental importance to find a single number which reflects some properties of the entire data. From that number, we can make inferences. For example, when you take a test, you may ask your teacher what the class average or the mean is. If the mean is high, you say it was an easy test. If the mean is low, you say that it was a hard test. The class average also gives you an idea of how your score compares with the rest of the class. The class average is an example of a number that measures central tendency. The median that we discussed in chapter 1 is also a measure of central tendency. The more commonly used measures of central tendency are the mean, the median, and the mode.

DEFINITION. The *mean* (or *average* or the *arithmetic mean*) of a set of numbers is the sum of all the numbers in the set divided by the number of elements in the set.

$$\text{Mean} = \frac{\text{Sum of the numbers in the set}}{\text{Number of elements in the set}}$$

For example, the mean of the numbers 2, 3, 4, 5, 6 is $\dfrac{2 + 3 + 4 + 5 + 6}{5} = 4$.

The Sigma Notation

In dealing with the mean, it is necessary to take the sum of a set of numbers. Therefore, it is convenient to use the sigma notation, Σ, defined by

$$\sum_{i=1}^{n} x_i = x_1 + x_2 + \ldots + x_n \qquad (1)$$

The sigma notation as described above contains

    (a)   A summation index ['i' in (1)],

    (b)   A range of values for the summation index $[i = 1, \ldots, n$ in (1)].

    (c)   An expression involving the summation index $[x_i]$.

To find the sum given by the sigma notation, one must substitute each possible value in the given expression for the summation index from its range, and take its sum. Consider the following:

$$\sum_{i=1}^{7} i = 1 + 2 + 3 + 4 + 5 + 6 + 7$$

$$\sum_{i=1}^{4} x^i = x^1 + x^2 + x^3 + x^4$$

$$\sum_{i=1}^{5} ax_i = ax_1 + ax_2 + ax_3 + ax_4 + ax_5$$

The choice of the index 'i' in the sigma notation is arbitrary; one can use any other letter. Thus, $\sum_{p=1}^{n} x_p$ is the same number as $\sum_{i=1}^{n} x_i$. We will use $\Sigma(x)$ to mean the sum of all possible values of x.

The mean of the n numbers $x_1, x_2, \ldots, x_n$ is denoted by $\overline{x}$. Thus,

$$\text{mean of } x_1, x_2, \ldots, x_n = \overline{x} = (x_1 + x_2 + \ldots + x_n) \div n$$

$$= \frac{x_1 + x_2 + \ldots + x_n}{n} \qquad (2)$$

$$= \frac{\Sigma(x_i)}{n} \qquad (3)$$

Examples 2.1

1. Find the mean of 10, 14, -8, 6, 7, -5, -3.

Solution. $\overline{x} = (10 + 14 - 8 + 6 + 7 - 5 - 3) \div 7 = 21 \div 7 = 3.$

2. The scores of 10 students in a test are: 76, 77, 48, 90, 85, 65, 67, 66, 71, 70. Find the mean.

Solution. mean $= (76 + 77 + 48 + 90 + 85 + 65 + 67 + 66 + 71 + 70) \div 10$

$$= \frac{715}{10} = 71.5.$$

Exercises   2.1

1. The scores of ten students in a test are 90, 87, 76, 92, 56, 72, 65, 63, 61, 71. Find the mean.

2. Find the mean rainfall during the first eight days of Spring if the amount of rainfall in inches during those days was

    1.9, 0.7, 2.7, 0.0, 0.0, 1.3, 1.01, 0.014.

3. The following are the lowest daily temperatures in centigrade of the first twelve days of Winter. Find the mean of these temperatures.

    -10, -8, 1, 4, -10, -10, 2, 3, -3, 11, 13, and 14.

4. Find the mean of the following test scores.

51	74	79	75	73	45	58	63
70	86	62	79	73	47	90	70
87	72	85	82	61	86	24	90

5. Find the mean of the following.

2	5	7	0	0	1	6	8	4	3
0	5	1	4	3	9	10	2	6	1
3	2	4	4	5	7	9	9	8	0

6. The following are the points scored by the University of Rhode Island's basketball team in the 1978-79 season. Find the average points scored by the team during that season.

    77 67 85 103 68 85 83 76 69 83 76 73 91 95
    86 71 67 99 48 68 69 79 84 77 98 75 75.

    16 3,138,152,202, 11 6,15 3,152,155,153,160,174,148,166, 95

For questions 7 - 10, use the responses to the questionnaire in Figure 1.1.

7. (a) Give the average height of male students in the class.
    (b) Give the average height of female students in the class.

8. (a) Give the average SAT math score of male students in the class.
    (b) Give the average SAT verbal score of male students in the class.

9. (a) Give the average SAT math score of female students in the class.
    (b) Give the average SAT verbal score of female students in the class.

10. (a) Give a reason for the difference between the average SAT math score of male students and the average SAT math score of female students in the class.

    (b) Give a reason for the difference between the average SAT verbal score of male students and the average SAT math score of female students.

    (c) Are there reasons to believe that the SAT discriminates against female?

11. Find each of the following: (a) $\sum_{i=1}^{6} i$ (b) $\sum_{i=1}^{6} i^2$

12. Find each of the following: (a) $\sum_{i=1}^{10} i$ (b) $\sum_{i=1}^{10} i^2$

13. Find each of the following: (a) $\sum_{i=1}^{5} x_i$ (b) $\sum_{i=1}^{5} x_i^2$

    (c) $\sum_{i=1}^{5} 3x_i$ (d) $\sum_{i=1}^{5} (x_i - 3)$

14. Find each of the following: (a) $\sum_{i=1}^{4} 5x_i$ (b) $\sum_{i=1}^{4} (x_i - 1)^2$

    (c) $\sum_{i=1}^{4} 2(x_i - 3)$ (d) $\sum_{i=1}^{6} (x_i - 7)$

The following example shows an application of the mean.

Examples 2.1

1. *(Mother's smoking linked to learning.)* Researchers from Cornell and Rochester University began a test with 400 pregnant women. Four years later, they gave the intelligence test to the children. It was found that children of women who smoked 10 or more cigarettes a day while pregnant had a mean score of 103 on the test. The mean score of children of non-smokers was 113. Even though factors such as maternal educationand social background influence scores in standard intelligence tests, the 10 point difference in the mean of test scores is conclusive enough to suggest that exposure to cigarette smoke in the womb may slow a child's ability to learn.

[Source: Providence Journal-Bulletin , March 7, 1994.]

The advantage of the mean is that it takes each value into account.

The disadvantages are:

(1) We need all the numbers to compute it. This may be an uphill task if there is a large number of elements in the data set.

(2) It is unduly affected by extremely low or high entries. In such a case, the mean may not be a good measure for the central tendency of the data.

For example, consider the professor who proudly declared to his class, "The class average on this test is 70. I am very happy with your performance. This is the best performance I have seen in years on this kind of test." It was found that the actual scores of the ten students of the class were 100, 100, 100, 59, 59, 55, 58, 58, 57, 54. In this case, the mean score or the class average of 70 does not reflect the fact that only three of the ten students passed the test. The mean also does not indicate that seven students got less than 60. The professor's claim that it was the best performance he had seen in years was not based on sound principle. In situations as this, the median or the mode may give a better indication of the population than the arithmetic mean.

Recall that the mode of the data is the number that has the greatest frequency exceeding the frequency of 1. The mode is an appropriate measure of central tendency if one number occurs quite frequently in the center of a distribution. Extremely high or low values do not affect the mode. It is easy to determine the mode of the data, if it exists. The disadvantage with the mode is that there may be more than one mode or there may not be a mode.

Relationship of the Mean, Median, and Mode

It is difficult to set fixed rules on whether to use the mean, the median, or the mode as a measure of central tendency. Very often the choice will depend on the nature of the data that is under consideration. However, the following guidelines may be helpful.

(1) If the frequency distribution is symmetrical, use whichever is easiest to find.

(2) If the mode appears at the center, use the mode.

(3) If the mode does not appear near the center, use the mean or the median.

(4) If the data contains some entry or entries which are extremely large or small compared to the rest, then use the median or the mode.

Graph of Frequency Distribution

The frequency distribution of data may be represented on a coordinate plane where the horizontal axis represents the numbers in the data and the vertical axis represents the frequency of each number in the data. A point in the coordinate plane will correspond to each number of the data and its frequency. By joining consecutive points by an arc or a line segment, we get a *graph of* the frequency distribution.

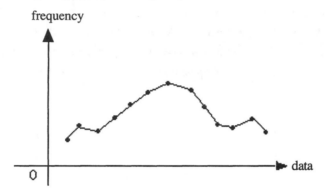

Figure 2.1. Frequency Graph

The graph of a frequency distribution is *symmetrical* if one side of the graph is a vertical reflection of the other side (Figure 2.2). In a symmetrical frequency distribution, the mean and the median are the same. Later on we will discuss a particular symmetrical distribution known as the normal distribution. As the name suggests, a large number of real world phenomena, like the scores in a test, heights of people in a college and so on have the normal distribution.

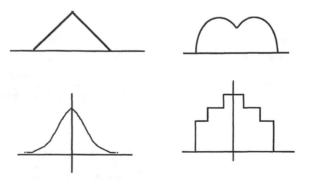

Figure 2.2. Examples of Symmetrical Frequency

Positively or Negatively skewed Data

The graph of a frequency distribution is *skewed* if it is not symmetrical. A frequency distribution is *positively skewed* if its graph has a longer tail to the right, that is, in the positive direction from the mean. This happens when the data contains some entries that are extremely high compared to the rest. For example, say you are in a class where the distribution of test scores is symmetrical and the mean is usually 60. Three bright students transfer to your class and they always score more than 95. Then the test scores of your class will become positively skewed because the high scores of the new students will pull the mean higher than it was before. Figure 2.3 shows what the frequency graph and the stem & leaf display of a positively skewed data may look like.

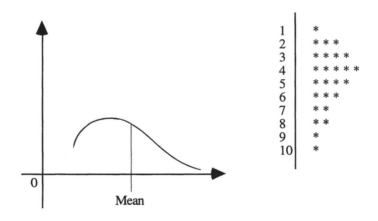

Figure 2.3. Positively Skewed Frequency

A frequency distribution is *negatively skewed* if its graph has a longer tail to the left, that is, in the negative direction from the mean. This happens when the data contains some entries that are extremely low in comparison with the rest. Figure 2.4 shows what the frequency graph and the stem & leaf display of a negatively skewed data may look like.

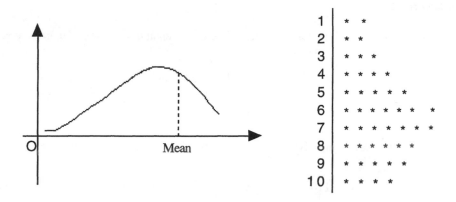

Figure 2.4. Negatively Skewed Frequency

Examples 2.3

1. The following are the scores of 12 students on a test. What should be used as a measure of central tendency and why?

    98, 100, 99, 67, 69, 71, 45, 55, 60, 58, 46, 55.

Solution. The mode 55, because it appears frequently (2 times) near the center. In this case, the mean, 68.58, and the median, 63.5, will reflect high scores even though half the class got 60 or less. The data is positively skewed.

2. The following are the scores of 13 students in a test. What should be used as a measure of central tendency for the data and why? Draw a frequency graph for the data, and determine whether it is symmetrical, positively skewed, or negatively skewed.

    90, 75, 80, 90, 60, 90, 70, 35, 65, 45, 85, 85, 75.

Solution. In this case, the mode 90 is too high and is not at the center. The median 75 is a good measure. The mean 72.69 is a good measure. This data is negatively skewed because the low scores 35 and 45 have brought the mean down. So, the mean 72.69 may be little low as a measure of central tendency.

Exercises 2.2

1. For each of the following data, find the mean and the mode, if there is any.
   - (a)   2, 9, 3, 7, 11, 12, 5
   - (b)   14, 10, 23, 14, 19, 18, 11, 8, 9, 12
   - (c)   20, 11, 35, 18, 22, 30, 15, 14, 11, 13
   - (d)   2, 3, 7, 15, 7, 11, 12, 5

2. The following give the life (in hours) of 20 light bulbs. Find the mean, median and mode.

111	110	103	81	86	110	100	80	107	108
85	101	104	88	101	105	99	75	87	101

   75 80 81 85 86 87 88 99 100, 101, 101, 101, 103 104 105 107 108 110 110 111.

3. The owner of a car keeps records of the mpg (milea per gallon of gasoline) that he gets. The following are the miles per gallon that he got under different driving conditions. What would be a good measure of mpg for his car?

   18 19 22 25 25 27 16 17 15 20

4. The following are the scores of 10 students on a test. Find the mean, median, and mode of the data. Which of these should be used as a measure of central tendency and why? Draw a graph of the frequency distribution and determine whether it is symmetrical, positively skewed, or negatively skewed.

   62 64 64 60 68 72 56 70 56 64
   4   5   6   3   8   10   2   9   1   7

5. The following are the scores of 20 students on a test. Find a suitable measure for central tendency and state why it is preferable to other measures of central tendency. Find, without drawing a graph of the frequency distribution, whether it is positively or negatively skewed.

64	22	72	64	56	28	74	80	80	56
59	72	80	64	64	100	99	72	68	74

6. The Brown and Sons company has 12 employees. The weekly salaries of the employees in hundreds are: 15 (Mr. Brown), 14 (for Mr. Brown's son), 13 (Mr. Brown's daughter), 12 (Mrs.Brown), and for the other employees, not related to Mr. Brown, the salaries, in hundreds, are 4, 3, 2, 2, 2.5, 2.5, 3, 5.Mr. Brown

1942

claims his employees are well paid because the average salary is $650 per week. Is Mr. Brown's claim misleading? Give reasons to support your answer.

7. *(Which school is better? Based on a true story).* The father of a girl decided to send her, against her wishes, to a nearby private school instead of the public school. He based his reason on the fact that the mean scores in the Scholastic Aptitude Tests for seniors in the private school exceeded that of the public school by about 50 points. Not knowing anything to counter her father's reasoning, the girl requested her elder brother Keith to find some ways to make their father change his mind. Responding to his sister's pleas, Keith went and checked the SAT scores of the seniors of two schools for several years. It was true that the mean for the private school seniors was usually higher than that of the public schools. He noticed a definite pattern both in scores in the verbal portion as well as in the math part. On the strength of his observations, Keith was able to persuade his father to change his mind. Study the scores of the seniors of the two schools as given below and determine what arguments Keith might have used to help his sister.

Private School (SAT Math Scores): 500, 480, 490, 540, 575, 630, 460, 460, 490, 490, 600, 630, 490, 480, 490, 480, 450, 490, 520, 490, 490, 605, 630, 480, 530, 450, 450, 490, 420, 460, 470, 610, 490, 500, 500.

Public School (SAT Math Scores): 330, 480, 480, 500, 280, 400, 500, 510, 530, 580, 410, 490, 490, 560, 580, 550, 410, 530, 560, 510, 480, 530, 300, 550, 550, 470, 500, 510, 500, 470, 430, 560, 550, 600, 400, 390, 450, 370, 510, 490.

8. Ask your teacher to read the scores of your class on a recent test. Find the mean, median, and mode, and state which of these will be a good measure of central tendency. Without drawing a graph of the frequency distribution, state whether the distribution is symmetrical, positively skewed, or negatively skewed.

9. Draw the graph of the frequency distribution for the following data and determine if it is symmetrical. 5 10 8 10 8 10 15 12 12.

10. Design a test to determine if the SAT discriminates against females.

2. THE MEAN OF GROUPED DATA

To determine the mean of grouped data,

Step 1. Draw a suitable portion of a number line and mark the points of division for the intervals.

Step 2. Find the midpoint of each interval after noting the points of division.

Step 3   Multiply the middle point of each interval by the frequency of that interval. The sum of these products divided by the total number of frequencies in the data is the mean. Thus

$$\text{Mean of Grouped Data} = \frac{\text{Sum of Midpoint} \times \text{Frequency}}{\text{Total Frequency}}.$$

Examples 2.4

1.   Find the mean of the data given in the following table [Note that in this case, the points of division are at 10, 20, 30, and 40].

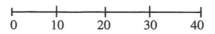

Interval	Frequency	Midpoint of Interval	Midpoint × Frequency
0 - 10	3	5	15
11 - 20	4	15	60
21 - 30	4	25	100
31 - 40	5	35	175
41 - 50	6	45	270

Total Frequency = 22   Sum of (Midpoint × Frequency) = 620

Mean = 620 ÷ 22 = 28. $\overline{18}$.

2.   The following data gives the grade distribution of 25 students in a test.

4 A's (90 - 100),   5 B's (80 - 89),   10 C's (70 - 79),

3 D's (60 - 69), and 3 F (below 60). What is the mean score for the test?

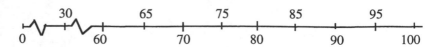

Solution.

Interval	Frequency	Midpoint of Interval	Midpoint × Frequency
0 - 59	3	30	90
60 - 69	3	65	195
70 - 79	10	75	750
80 - 89	5	85	425
90 - 100	4	95	380

Total Frequency = 25　　　　Sum of (Midpoint × Frequency) = 1840

Mean = $\dfrac{1840}{25}$ = 73.6.

Using the sigma notation, we can give a formula for the mean of grouped data in the following manner. If x_1 is the midpoint of the first interval that has frequency f_1, x_2 is the midpoint of the second interval that has frequency f_2, and so on. The formula to determine the mean $\overline{x}$ is

$$\overline{x} = \frac{\Sigma(xf)}{\Sigma(f)} = \frac{x_1f_1 + x_2f_2 + \ldots + x_nf_n}{f_1 + f_2 + \ldots + f_n} \tag{4}$$

The interval that has the greatest frequency gives some central tendency of grouped data. This interval is the *modal class*. The modal class of the data in 1 of Examples 2.4 is 41- 50.

Exercises 2.3

1.　Find the mean and the modal class of each of the following data.

(a)

Interval	Frequency
0 - 10	3
11 - 20	6
21 - 30	14
31 - 40	9
41 - 50	5

(b)

Interval	Frequency
-10 - (-5)	2
-4 - 0	3
1 - 5	8
6 - 10	12

2.   Find the mean and state the modal class of each of the following data.
      (a)

      (b)

Interval	Frequency
0 - 4	3
5 - 8	7
9 - 12	10
13 - 16	10
17 - 20	6
21 - 24	2

Interval	Frequency
-8 - (-4)	5
-3 - 0	6
1 - 4	1
5 - 8	3

3.   The following data gives the grade distribution of 30 students in a test.

         5 A's (90 - 100),   7 B's (80 - 89),   12 C's (70 - 79),

         5 D's (60 - 69), and 1 F (below 60). What is the mean score for the test?

4.   The following data gives the grade distribution of 50 students in a test.

         8 A's (90 - 100),   10 B's (80 - 89),   18 C's (70 - 79),

         8 D's (60 - 69), and 6 F (below 60). What is the mean score for the test?

5.   The lowest temperatures (in degree Fahrenheit) for January, 1994 in a Canadian town were between -10 and -5 for 10 days, between -5 and 0 for 12 days, between 0 and 5 for 6 days and between 5 and 10 for the remaining days. Find the mean of the lowest temperatures in the city for the month of January.

6.   The lowest temperatures (in degree Fahrenheit) for January, 1995 in the Canadian town of problem 5, were between -10 and -5 for 8 days, between -5 and 0 for 15 days, between 0 and 5 for 6 days and between 5 and 10 for the remaining days. Find the mean of the lowest temperatures in the city for the month of January. Which year was colder in January for the city? [See problem 5.]

7.   Group the SAT math scores of the private school seniors into intervals of 50 and determine the modal class (see problem 7 of Exercises 2.2).
      Private School (SAT Math Scores): 500, 480, 490, 540, 575, 630, 460, 460, 490, 490, 600, 630, 490, 480, 490, 480, 450, 490, 520, 490, 490, 605, 630, 480, 530, 450, 450, 490, 420, 460, 470, 610, 490, 500, 500.

8.   Group the SAT math scores of the public school seniors into intervals of 50 and determine the modal class (see problem 7 of Exercises 2.2).
      Public School (SAT Math Scores): 330, 480, 480, 500, 280, 400, 500, 510, 530, 580, 410, 490, 490, 560, 580, 550, 410, 530, 560, 510, 480, 530, 300, 550, 550, 470, 500, 510, 500, 470, 430, 560, 550, 600, 400, 390, 450, 370, 510, 390.

3. MEASURES OF VARIATION

Besides the central tendency, one can also analyze variations in given data. For example, in a test you receive a score of 70 and the teacher tells you that the class average is 60. What can you say about your performance? Are you a superior, a very superior, or just an average student in the class? Below are three different distributions, each of which has an average of 60 in a class of 10.

I. 50, 54, 55, 57, 66, 58, 60, 68, 62, 70
II. 30, 30, 45, 48, 70, 74, 75, 75, 76, 77
III. 40, 45, 50, 50, 60, 70, 70, 70, 70, 75

Each of these reflects differently about the score 70. If the distribution is like that in I, then you are definitely the best student in the class. If the distribution is like that in II, then you are just about an average student. If the distribution is like in III, then you are one of the better students in the class.

We can analyze distributions by measuring how the entries in a given set of data vary. There are several standard measures of such variations, also called *dispersions*. They are the range, the mean deviation, the variance and the standard deviation.

DEFINITION. The difference between the largest and the smallest numbers in a set is the *range* of the numbers (or data).

For example, in I the range is 70 - 50 = 20. In II, the range is 77 - 30 = 47 and in III, the range is 75 - 40 = 35.

DEFINITION. In a given set of numbers, the *deviation* of a number in the set is equal to the number minus the mean.

To find the deviation of a number in the set, first find the mean of the numbers, then subtract the mean from the number. Thus in data I, the deviation of 70 is 70 - 60 = 10 and the deviation of 55 is 55 - 60 = -5. The deviation of a number is positive if the number is higher than the mean and it is negative if the number is smaller than the mean.

Often, we need to consider the absolute value of deviation. The absolute value of a number is indicated by putting the number in between two vertical lines as in | -5 |. The absolute value of a number is always positive. The absolute value of a negative number is it's opposite positive. Thus | -5 | = 5. The absolute value of a positive number or zero is the number itself.

DEFINITION. The *mean deviation* of a data set is the mean of the absolute values of the deviations of the numbers in the data.

To find the mean deviation of a given data use the following steps.

Step-1. Find the mean.

Step-2. Tabulate the deviation and the absolute value of deviations.

Step-3. Find the mean of the absolute values of deviations.

We illustrate these steps by finding the mean deviation of

data I: 50, 54, 55, 57, 66, 58, 60, 68, 62, 70.

Score	Deviation	|Deviation|
50	50 - 60 = -10	10
54	54 - 60 = -6	6
55	55 - 60 = -5	5
57	57 - 60 = -3	3
58	58 - 60 = -2	2
60	60 - 60 = 0	0
62	62 - 60 = 2	2
66	66 - 60 = 6	6
68	68 - 60 = -8	8
70	70 - 60 = 10	10
Total = 60		52

$$\text{Mean} = \frac{600}{10} = 60 \qquad\qquad \text{Mean Deviation} = \frac{52}{10} = 5.2$$

The |deviation| signifies that we are taking positive values only.

The general formula for mean deviation may be given in the following manner. Let $x_1, x_2, \ldots, x_n$ be the numbers in the data. Let $\overline{x}$ be the mean. The deviation of x_1 is then $x_1 - \overline{x}$, that of x_2 is $x_2 - \overline{x}$, and so on. The formula for the mean deviation is

$$\text{Mean deviation} = \frac{|x_1 - \overline{x}| + |x_2 - \overline{x}| + \ldots + |x_n - \overline{x}|}{n}, \qquad (5)$$

where the notation $|x_i - \overline{x}|$ signifies that only the absolute value of $x_i - \overline{x}$ is to be considered.

The mean deviation indicates how the scores are spread about the mean. In case of test scores, the scores that are within one deviation of the mean are taken as average - these are the scores between (Mean - Mean deviation) and (Mean + Mean deviation).

Examples 2.5

1.   Find the mean and the mean deviation of

    data II: 30, 30, 45, 48, 70, 74, 75, 75, 76, 77.

Solution. We organize the data in the following table:

Score	Deviation	|Deviation|
30	-30	30
30	-30	30
45	-15	15
48	-12	12
70	10	10
74	14	14
75	15	15
75	15	15
76	16	16
77	17	17
Total = 600		174

$$\text{Mean} = \frac{600}{10} = 60 \qquad\qquad \text{Mean Deviation} = \frac{174}{10} = 17.4.$$

The \|deviation\| signifies that we are taking positive values only.

2.   If a number occurs more than once, then we may multiply it by its frequency instead of using the number again. Find the mean and the mean deviation of

    data III: 40, 45, 50, 50, 60, 70, 70, 70, 70, 75.

Score	Frequency	Score × Frequency	|Deviation|	|Deviation| × Frequency
40	1	40	20	20
45	1	45	15	15
50	2	100	10	20
60	1	60	0	0
70	4	280	10	40
75	1	75	15	15
Total	10	600		110

$$\text{Mean} = \frac{600}{10} = 60 \qquad\qquad \text{Mean Deviation} = \frac{110}{10} = 11.$$

In the case of I, the score of 70 is well beyond the mean deviation of 5.2. This reflects a very good score. In the case of II, the score of 70 is well within one mean deviation indicating that there are perhaps many students who scored more than 70. In the case of III, the score of 70 is just about one mean deviation from the mean, suggesting that it is just about an average score.

The mean deviation works quite well for most situations, is easy to determine without the help of a calculator or a square root table, and is quite satisfactory. Two other measures of dispersion are very popular. We shall discuss them now.

DEFINITION. The *variance* of a set of numbers is the mean of the squares of the deviations.

The variance of the numbers are $x_1, x_2, \ldots, x_n$, is written as $\text{Var}(x_1, x_2, \ldots, x_n)$. So,

$$\text{Var}(x_1, x_2, \ldots, x_n) = \frac{(x_1 - \bar{x})^2 + (x_2 - \bar{x})^2 + \ldots + (x_n - \bar{x})^2}{n}, \qquad (6)$$

where $\bar{x}$ is the mean of the numbers.

To find the variance of a given data use the following steps.

Step-1. Find the mean.

Step-2. Tabulate the deviation and the squares of deviations.

Step-3. Find the mean of the squares of deviations.

Thus to find the variance of: 50, 54, 55, 57, 66, 58, 60, 68, 62, 70

Score	Deviation	(Deviation)^2
50	-10	100
54	-6	36
55	-5	25
57	-3	9
58	-2	4
60	0	0
62	2	4
66	6	36
68	8	64
70	10	100
Total = 600		378

$$\text{Mean} = 60, \quad \text{Variance} = \frac{378}{10} = 37.8.$$

DEFINITION. The the *standard deviation* σ of a set of numbers is the square root of the variance of the numbers.

$$\text{Standard Deviation} = \sigma \text{ of } x_1, x_2, \ldots, x_n = \sqrt{\text{Var}(x_1, x_2, \ldots, x_n)}$$

$$= \sqrt{\frac{(x_1 - \overline{x})^2 + (x_2 - \overline{x})^2 + \ldots + (x_n - \overline{x})^2}{n}} \qquad (7)$$

Notice that $\overline{x}$ is the mean of the numbers.

Note 1. The standard deviation is very important in probability theory. The main disadvantage of standard deviation is that it involves square roots of numbers, which are not always easy to determine. However, calculators have virtually wiped out this disadvantage.

Note 2. One advantage of standard deviation is that it takes the following simple algebraic form.

$$\sigma = \frac{1}{n} \sqrt{n(x_1^2 + x_2^2 + \ldots + x_n^2) - (x_1 + x_2 + \ldots + x_n)^2} \ . \qquad (8)$$

A small standard (or mean) deviation indicates that the numbers cluster close to the mean and a relatively higher standard (or mean) deviation indicates that the numbers are widely scattered from the mean.

In the case of a distribution that is normal or approximately normal, about 68% of the entries in the distribution lie within one standard deviation of the mean, about 95% are within two standard deviations of the mean, and more than 99% are within three standard deviations of the mean. This is accepted as an empirical rule (see boxed material).

Empirical Rule

For a distribution that is approximately normal

the interval $\overline{x} - \sigma$ to $\overline{x} + \sigma$ contains 68% of the entries,

the interval $\overline{x} - 2\sigma$ to $\overline{x} + 2\sigma$ contains 95% of the entries,

the interval $\overline{x} - 3\sigma$ to $\overline{x} + 3\sigma$ contains almost all the entries,

where $\overline{x}$ is the mean.

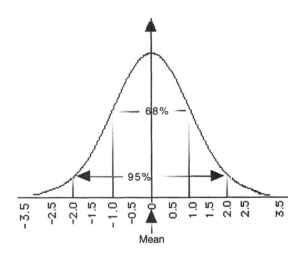

Figure 2.5

Therefore, in a test, the scores that are between (Mean $- \sigma$) and (Mean $+ \sigma$) are usually taken as average. Scores less than (Mean $- \sigma$) are below average and scores less than (Mean $- 2\sigma$) are poor. Similarly, scores more than (Mean $+ \sigma$) are better than average and scores above (Mean $+ 2\sigma$) are excellent.

Examples 2.6

1. Find the mean deviation, the variance, and the standard deviation of the numbers -1, -2, 4, 3, 5, 8, and 11.

Solution. Mean = 4

Number	Deviation	Square of deviation
-1	(-1) - 4 = -5	25
-2	(-2) - 4 = -6	36
4	4 - 4 = 0	0
3	3 - 4 = -1	1
5	5 - 4 = 1	1
8	8 - 4 = 4	16
11	11 - 4 = 7	49

Total   128

Mean deviation = 24/7 = 3.42.

Variance $= \dfrac{128}{7} = 18.285$.   Standard deviation $= \sqrt{18.285} = 4.276$.

2. The scores in a test of 10 students are:

50, 55, 57, 57, 66, 57, 60, 68, 60, 70.

Find the variance and the standard deviation.

Solution.

Score	Frequency	Deviation	(Deviation)^2	(Deviation)^2 × Frequency
50	1	-10	100	100
55	1	-5	25	25
57	3	-3	9	27
60	2	0	0	0
66	1	6	36	36
68	1	8	64	64
70	1	10	100	100
Total	10			352

Mean = 60,   variance = 35.2,

Standard deviation $= \sigma = \sqrt{\dfrac{352}{10}} = \sqrt{35.2} = 5.93$.

3. For the data in 2, find each of the following:
   (a)   The test scores that are average (within one standard deviation of the mean).
   (b)   The test scores that are better than the average (more than Mean + σ.).
   (c)   The test scores that are excellent (more than Mean + 2σ).
   (d)   The test scores that are below average (less than Mean - σ.)
   (e)   The test scores that are poor (less than Mean - 2σ.)

Solution.   (a) 55, 57, 60   (b) 66, 68, 70   (c) None

(d) 50   (e) None

Exercises 2.4

1. Find the range, the mean and the mean deviation for each of the following:
   (a)   12, 11, 17, 7, 4, 9
   (b)   7, 8, 8, 12, 10
   (c)   4, 5, 6, 8, 6, 6, 11, 12, 12, 10
   (d)   12, 13, 8, 8, 6, 13, 13, 13, 13
   (e)   5, 5, 5, 7, 7, 3, 4, 4, 8, 8, 1, 3, 9, 9, 9, 9

2. Find the variance and the standard deviation of each of the data in problem 1.

3. The scores in a test of 10 students are: 40, 45, 50, 50, 60, 70, 70, 70, 70, 75. Find the mean deviation, the variance and the standard deviation.

4. The scores in a test of 12 students are: 30, 45, 55, 58, 60, 70, 70, 75, 80, 85, 90, 90. Find the mean deviation, the variance and the standard deviation.

5. For the data in problem 3, find each of the following:
   (a)   The test scores that are within one standard deviation of the mean.
   (b)   The test scores that are above average.
   (c)   The test scores that are excellent.

6. For the data in problem 4, find each of the following:
   (a)   The test scores that are within one standard deviation of the mean.
   (b)   The test scores that are below average.
   (c)   The test scores that are poor.

7. For the data: 5, 9, 23, 14, 12, 14, 25, 10, 18, 10, find each of the following:
   (a)   The standard deviation.
   (b)   The numbers that are within one standard deviation of the mean.

(c)   The numbers that are within two standard deviations of the mean.

8.   The class average on a test is 70, the standard deviation is 2.3, and your teacher informs you that the distribution is fairly symmetrical.

(a) If your score is 80, can you conclude that you are one of the top students?

(b) If your score is 68, can you conclude that you are one of the inferior students?

9.   The following are the points scored by the eight players of a basketball team. Rate the performance of each player as poor, average, above average or excellent by using the standard deviation.

Player	A	B	C	D	E	F	G	H	I	J
Scores	13	19	7	22	10	12	9	11	12	5

10.   The following are the points scored by the eight players of a basketball team. Rate the performance of each player as poor, average, above average or excellent by using the standard deviation.

Player	A	B	C	D	E	F	G	H	I	J
Scores	22	4	14	10	11	11	9	2	12	15

11.   A college basketball coach looking for recruits had to decide from three prospects who were equal in all respects. The coach decided to pick the one who had the better scoring ability. The number of points scored by each of the prospects in 10 games are given below. Which player should the coach select and why?

A: 20, 24, 36, 4, 30, 5, 21, 15, 25, 19

B: 1, 7, 31, 29, 5, 37, 27, 13, 11, 39

C: 20, 22, 20, 20, 18, 22, 18, 20, 18, 22

12.   The following are the points scored by two basketball players in one season. Determine which player is more consistent.

A :   10, 12, 13, 12, 14, 20, 25, 14

B :   11, 15, 5, 21, 9, 35, 14, 10

13.   Your teacher gives you the test scores of the students in your class. State how you may judge your performance in the test.

4. CORRELATION COEFFICIENT

The mean, median, standard deviation, etc., are measures used to study the behavior of one set of numbers. Now we wish to discuss how to correlate the behavior of two sets of numbers. It will be convenient to give the two sets names like X and Y. For example, in the following table,

(handwritten: $X = 8$)

(handwritten above table: $\bar{X} = 8$　-2　-2　-2　-1　0　0　1　2　2　2$)

(handwritten left: $X = 8$　$\bar{Y} = 70$)

X (Homework)	6	6	6	7	8	8	9	10	10	10
Y (Test)	40	50	80	60	85	90	90	60	50	95

X is the homework scores of ten students and Y is the test scores of the same students. Here X and Y are both *variables* in the sense that they each represent a set of numbers. We can find the mean and the mean deviation for both X and Y, but this information would not tell us about any relationship between the two. At the first glance it seems that in some cases a person with high homework score has low test scores and some with low homework scores have high test scores. Just by comparing, we cannot say whether the test scores increase with homework scores or not. Therefore, we wish to devise a method that will measure the degree of association of the two variables.

It is quite natural to think that your test scores are likely to increase if you spend more time doing your homework. Similarly, if you do not take your homework seriously you are likely to score less in your homework and your test score is likely to decrease. In other words, the variables either both increase or both decrease. This is a case of direct correlation.

Two variables may be such (see problem 2 of Examples 2.7) that if one increases then the other decreases and vice versa. In this case, there is an inverse correlation.

Correlation between two variables X and Y is measured by the correlation coefficient r. The steps to determine r are as follows.

Step 1.　Find the mean $\bar{x}$ of X.

Step 2.　For each value x_i of X, tabulate the deviation $(x_i - \bar{x})$. [This is the actual difference with the mean and therefore, it is positive if x_i is higher than the mean, and is negative if x_i is smaller than the mean.]

Step 3.　Find the mean $\bar{y}$ of Y.

Step 4.　For each value y_i of Y tabulate $(y_i - \bar{y})$ as in step 2.

Step 5.   Tabulate the product of the deviations of corresponding values of X and Y. Find the sum.

Step 6.   On separate columns tabulate the (deviation)^2 for X and for Y. Find the respective sums.

Step 7.   Determine the correlation coefficient r by using the following formula:

$$r = \frac{\text{Sum of the product of deviations}}{\sqrt{(\text{Sum of (deviations)}^2 \text{ for X}) \times (\text{Sum of (deviations)}^2 \text{ for Y})}} \qquad (9)$$

Using the summation notation, formula (9) may be expressed in the following form:

$$r = \frac{\Sigma(x_i - \bar{x})(y_i - \bar{y})}{\sqrt{(\Sigma(x_i - \bar{x})^2)(\Sigma(y_i - \bar{y})^2)}} \qquad (10)$$

Notes. (1)   The range of values for the correlation coefficient r is -1 to 1.

(2)   If r = 0, then there is no correlation between the variables X and Y.

(3)   If r is positive, then there is a *direct correlation* between X and Y. In this case, if one variable increases the other increases. The correlation is *high* if r is high and is *perfect* if r = 1.

(4)   If r is negative, then there is *inverse correlation*. That is if one increases the other decreases and vice versa.

The good thing about correlation is that we do not need any new data. The bad thing about it is that we have to use both squaring and square roots.

Examples 2.7

1.   The following table gives the homework scores and the scores in the first test of ten students last semester.

X (Homework)	6	6	6	7	8	8	9	10	10	10
Y (Test)	40	50	80	60	85	90	90	60	50	95

(a)   Is it likely that there is a correlation between the homework scores and the test scores? Can it be a direct or inverse correlation?

(b) Determine the correlation coefficient between the homework scores and the test scores.

Solution.

(a) It is likely to be a direct correlation because in most cases the test scores are higher for the students with higher homework scores.

(b) Let X be the variable representing homework scores and Y be the variable representing test scores. Then we get the following:

X	deviation	(deviation)²	Y	deviation	(deviation)²	Product of deviations
6	-2	4	40	-30	900	60
6	-2	4	50	-20	400	40
6	-2	4	80	10	100	-20
7	-1	1	60	-10	100	10
8	0	0	85	15	225	0
8	0	0	90	20	400	0
9	1	1	90	20	400	20
10	2	4	60	-10	100	-20
10	2	4	50	-20	400	-40
10	2	4	95	25	625	50
Total 80		Total 26	Total 700		Total 3,650	Total 100

$$\overline{x} = \frac{80}{10} = 8, \qquad\qquad \overline{y} = \frac{700}{10} = 70$$

Sum of the product of deviations = 100,   Sum of (deviations)² for X = 26,
Sum of (deviations)² for Y = 3,650.

$$\text{So,} \quad r = \frac{100}{\sqrt{26 \times 3650}} = \frac{100}{\sqrt{94900}} = .32.$$

Conclusion. The correlation coefficient .32 is not very high . This may be partly because two students who had perfect scores in homework did poorly on the test, giving a negative product. In such cases, the teacher is likely to suspect that those students perhaps did not really do their homework on their own.

2. (*An example of inverse correlation.*) The following table gives the hours per week that ten students spend watching TV and their love for math (on a scale of 0 - 10, a score of 0 means they do not like math and a score of 10 indicates a very high degree of liking for math).

X (love for math)	7	6	0	7	0	4	2	8	2	4
Y (TV hours)	10	15	15	5	30	20	16	6	15	8

(a) Is there any correlation between the time a student spends watching TV and the student's love for math?

(b) Determine the correlation coefficient between the time a student spends watching TV and the student's love for math.

Solution. (a) We get the following table.

X			Y			
X	deviation	(deviation)²	Y	deviation	(deviation)²	Product of deviations
7	3	9	10	-4	16	-12
6	2	4	15	1	1	2
0	-4	16	15	1	1	-4
7	3	9	5	-9	81	-27
0	-4	16	30	16	256	-64
4	0	0	20	6	36	0
2	-2	4	16	2	4	-4
8	4	16	6	-8	64	-32
2	-2	4	15	1	1	-2
4	0	0	8	-6	36	0
Total 40		78	140		496	-143

$$\overline{x} = \frac{40}{10} = 4, \qquad \overline{y} = \frac{140}{10} = 14$$

Sum of the product of deviations $= -143$, Sum of (deviations)² for X $= 78$, Sum of (deviations)² for Y $= 496$.

So, $r = \dfrac{-143}{\sqrt{78 \times 496}} = \dfrac{-143}{\sqrt{38688}} = -.73.$

Conclusion. This negative correlation may be because learning mathematics requires active participation of the students whereas watching TV is known to make people passive learners. In any case a high degree of correlation does not necessarily mean that one variable causes the other. It just indicates a possible link. To find the truth we need careful studies and analysis.

Exercises 2.5

1. It is now accepted that smoking causes cancer. Is there any correlation between the number of cigarettes a person smokes per day and the risk of becoming a cancer victim? Will the correlation coefficient in this case be positive, negative or zero?

2. The following table gives the values of two variables X and Y. Is there any correlation between X and Y? Determine the correlation coefficient.

X	2	4	6	8	10
Y	3	7	13	15	17

3. The following table gives the values of two variables X and Y. Is there any correlation between X and Y? Determine the correlation coefficient.

X	1	4	5	2	3
Y	10	4	2	8	6

4. Let the following table give the values of two variables X and Y. Is there any correlation between X and Y? Determine the correlation coefficient.

X	3	4	2	7	8	9	11	12
Y	8	7	6	5	5	4	3	2

5. The following table gives the number of cigarettes a person smokes per day and his or her risk of becoming a cancer victim (the data is hypothetical).

# of cigarettes	5	8	10	13	15	21	23	25
risk of cancer	.15	.25	.35	.45	.5	.6	.8	.9

   Determine the correlation coefficient between the number of cigarettes a person smokes and the risk of the person becoming a cancer victim.

6. The following table gives the values of two variables X and Y. Is there any correlation between X and Y? Determine the correlation coefficient.

X	25	22	18	15	12	10	8	5	3	2
Y	4	5	7	8	9	10	12	14	15	16

7. The following table gives the number of days ten students attended math classes during the first four weeks of last semester, and the respective test scores in the test that followed.

attendance	12	12	11	10	10	9	8	7	6	5
test scores	70	90	100	75	80	70	55	50	60	40

(a) Is there any correlation between the number of days the students attended classes and their test scores? Will the correlation coefficient be positive or negative?

(b) Determine the correlation coefficient between students' attendance records and their test scores last semester.

(c) Is it a case of direct, inverse, or no correlation?

8. The following table gives the SAT math and SAT verbal scores of eight male students.

verbal	410	460	490	510	510	520	580	600
math	410	500	510	420	490	640	210	500

(a) Is there any correlation between the SAT math scores and the SAT verbal scores of the male students? Can it be a direct or an inverse correlation?

(b) Determine the correlation coefficient between the SAT math scores and the SAT verbal scores of the eight male students last semester.

9. The following table gives the SAT math and SAT verbal scores of eight female students.

verbal	340	370	390	400	430	450	540	600
math	460	470	440	200	450	500	360	320

(a) Is there any correlation between the SAT math scores and the SAT verbal scores of female students? Can it be a positive or a negative correlation?

(b) Determine the correlation coefficient between the SAT math scores and the SAT verbal scores of ten female students last semester.

10. *Men who can 'hold' liquor risk alcoholism* - this conclusion is based on a study carried on over a period of ten years. The study related the alcohol responses of men who were around age 20 when tested, to their risk of alcoholism by the time they were contacted an average of 9.3 years later. The testing used two alcohol doses that produced the same blood alcohol concentration as drinking about three and five beers within 10 minutes. The larger doses would get somebody legally drunk in most states. Is there any correlation between how high a person feels the first time he or she tries drinking and the risk of becoming an alcoholic? Can it be a direct or inverse correlation? Will the correlation coefficient be positive or negative? What message can you draw from this study?

[Source: *American Journal of Psychiatry*, Jan., 1994.]

11. *(A New Approach to Population Control.)* Experts at the United Nations have concluded that family planning alone will not adequately curb population growth and that education and prenatal care are essential to success. This conclusion is based on data in Figure 2.6. **Note.** The female education index is based on measurements of the enrollment of girls in primary and secondary school, differences in enrollment between girls and boys and the average level of schooling attained by women. [Source. *The New York Times*, April 13, 1994].

Country	Female Education Index	Infant Mortality Rate (deaths/1000births)	Fertility Rate Number of Children
Japan	95.7	5	1.6
France	99.7	7	1.8
US	97.7	9	1.9
Namibia	72.2	100	5.9
China	67.0	29	2.5
Thailand	66.5	27	2.5
Zimbabwe	66.2	49	4.9
Kenya	57.0	67	6.5
India	50.4	92	4.0
Rwanda	50.2	120	8.3

Figure 2.6

(a) Do countries with a higher female education tend to have lower or higher infant mortality rate?

(b) Find the correlation coefficient between the female education index and the mortality rate of the countries in the data.

(c) Do countries with a higher female education tend to have lower or higher fertility rate?

(d) Find the correlation coefficient between the female education index and the fertility rate of the countries in the data.

CHAPTER TEST 2

1. The dollar amount spent by eight students on a Saturday night are 5, 9, 23, 14, 25, 10, 18, 20. For the data, find the following.

    (a) the mean,   (b) the median,   (c) the mode,   (d) the range.

2. The scores ten students on a test are   91, 78, 77, 94, 75, 81, 87, 88, 89, 90. What is the mean deviation for the data?

3. Find the standard deviation for the data in problem 2.

4. Complete the following frequency table for grouped data and compute the mean for the data shown.

Interval	Midpoint	Frequencies
0 - 10	5	5
11 - 20		3
21 - 30		4
- 40		2
41 - 50		6

5. The following are the points scored by the eight players of a basketball team. Rate the performance of each player as poor, average, above average or excellent by using the standard deviation.

Player	A	B	C	D	E	F	G	H
Scores	13	11	7	20	15	12	4	6

6. The following data gives the grade distribution of 40 students in a test.

    5 A's (90 - 100),   8 B's (80 - 89),   15 C's (70 - 79),
    10 D's (60 - 69), and   2 F's (below 60).

    (a) Estimate the mean score in the test.

    (b) Give the relative frequency of each grade.

7. Each of the following gives the homework scores of a student. For each, determine if the frequency graph will be symmetrical, positively skewed or negatively skewed.

    (a) 2, 8, 8, 7, 6, 7, 4, 8, 8, 9, 10, 10, 9, 8, 7, 6, 7, 6, 4, 5, 8, 9

    (b) 7, 8, 6, 6, 9, 5, 10, 4, 8, 4, 8, 6, 10, 9, 5, 6, 5, 9, 8, 7.

8. The mpg (miles per gallon) that a person recorded for his new car under different driving conditions were 19, 23, 22, 27, 24, 27, 26, 27, 35, 30.

    Construct a stem and leaf display for the data. What would be a good measure of mpg for his car?

9. For each of the following, state if you would expect a direct correlation, inverse correlation or no correlation.

    (a) Number of cigarettes women smoke while pregnant and the IQ of children.

    (b) The cups of coffee you drink in a day and the hours of sleep you get.

    (c) Hours you spend watching TV and your test scores.

    (d) Waistline of persons and their weights.

    (e) Heights of students and their SAT scores.

    (f) Female education index and the fertility rate of women of a country.

10. Let the following table give the values of two variables X and Y. Is there any correlation between X and Y? Determine the correlation coefficient.

X	1	2	3	4	5	7	8	9	10	11
Y	4	5	7	8	9	10	12	14	15	16

Chapter 3

SET LANGUAGE

1. "YES - NO" QUESTIONNAIRE

One convenient way of collecting data is with a Yes-No questionnaire. Written questions that are to be answered by 'Yes' or 'No' are handed to the target population. Answers are then collected and organized into a set diagram. For example, a survey was made in a math class. Students were asked to complete a set of Yes-No questionnaires. The first two questions were:

1. Do you like a class at 8 O'clock in the morning?   Yes   No
2. Should students be allowed to eat food in the classroom?   Yes   No

The table in Figure 3.2 gives the answers of 25 students. Clearly, it is difficult to make sense out of this data. But if we use sets and put them on a set diagram, we will be able to see a lot of things. To do so, first we have to separate the students into sets according to the answers. We usually name a set with an upper case letter as follows.

U = The set of all students surveyed. We call this the *universal set* of our survey.

A = The set of all students who like an 8 o'clock class. These are the students who answered 'Yes' to the first question.

$\overline{A}$ (complement of A) = The set of all students who do not like an 8 o'clock class. These are the students who answered 'No' to the first question.

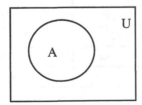
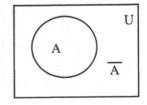

Figure 3.1

Venn diagrams. We represent the universal set U by a rectangular region and indicate A by a closed region inside the rectangular region [Figure 3.1]. Then, a point inside the closed curve represents a student of the set A. These kind of diagrams for sets are known as Venn diagrams. Then $\overline{A}$ will be represented by the region in U that is outside A.

Students	Question 1	Question 2
1.	Yes	Yes
2.	No	Yes
3.	Yes	No
4.	No	No
5.	No	No
6.	Yes	No
7.	Yes	Yes
8.	Yes	No
9.	Yes	No
10.	Yes	No
11.	No	No
12.	Yes	Yes
13.	No	Yes
14.	No	Yes
15.	Yes	No
16.	Yes	No
17.	No	No
18.	No	Yes
19.	No	Yes
20.	Yes	Yes
21.	Yes	No
22.	Yes	No
23.	Yes	No
24.	No	No
25.	No	Yes

Figure 3.2

Similarly, let

  B = The set of all students who think students should be allowed to eat in classrooms. These are the students who answered 'Yes' to the second question.

  $\overline{B}$ = The set of all students who do not think that students should be allowed to eat in classrooms. These are the students who answered 'No' to the second question.

  We do the same thing for B as we did for A. In reality, we do not have to make two separate diagrams for each of the sets A and B because we can put them on the same diagram [Figure 3.3]. Then we will have four distinct regions of U, which are described in set languages as follows:

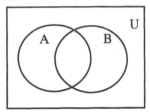 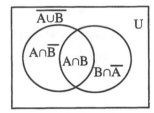

Figure 3.3

A ∩ B (A intersection B) = This is the region common to both A and B. This corresponds to the students who answered 'Yes' to both question.

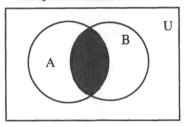 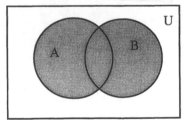

A ∩ B (shaded region)   A ∪ B (shaded region)

Figure 3.4

A ∪ B (A union B) = This is the region which is either in A, in B or in both. This corresponds to the students who answered 'Yes' to one or both of the two questions.

$\overline{A \cup B}$ (complement of A union B) = This is the region which is neither in A nor in B. This corresponds to the students who answered 'No' to both question.

In addition, we have the following sets:

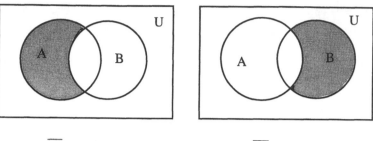

A ∩ $\overline{B}$ (shaded region)   B ∩ $\overline{A}$ (shaded region)

Figure 3.5

A ∩ $\overline{B}$ (A intersection complement of B) = This is the region of A which is not in B.

This corresponds to the students who answered 'Yes' to the first question and 'No' to the second question.

B ∩ $\overline{A}$ (B intersection complement of A) = This is the region of B which is not in A.

This corresponds to the students who answered 'Yes' to the second question and 'No' to the first question.

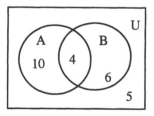

Figure 3.6

If we tally the number of students in each of these regions according to his or her answer, we will get the whole picture as shown in Figure 3.6.

Of course, we are interested in counting the number of students in each region. To do so, we will have to develop some symbols and formulas for counting. One important symbol is n(A) which stands for the *number of elements* in the region A. Remember that any time we write n( ) we mean a number.

So, n(A ∩ B) means the number of elements in A ∩ B. To find n(A ∩ B), count the number of elements that are common to A and B.

When solving counting problems involving two sets, say A and B, it is important to figure out $n(A \cap B)$ first. Then to find $n(A \cup B)$, use the following formula,

$$n(A \cup B) \quad = \quad n(A) + n(B) - n(A \cap B) \qquad (1)$$

Similarly, $n(\overline{A})$ means the number elements that are not in A. The formula to determine $n(\overline{A})$ is as follows.

$$n(\overline{A}) \quad = \quad n(U) - n(A) \qquad (2)$$

Therefore, to find $n(\overline{A})$, simply subtract the number of elements in A from the total number of elements in the universal set. In the same way, we get

$$n(\overline{A \cup B}) \quad = \quad n(U) - n(A \cup B) \qquad (3)$$

Since $A \cap \overline{B}$ is the set of elements that are in A but outside B, we get

$$n(A \cap \overline{B}) \quad = \quad n(A) - n(A \cap B) \qquad (4)$$

Similarly,

$$n(B \cap \overline{A}) = n(B) - n(A \cap B) \qquad (5)$$

We can now answers questions such as:

(a) How many students do not like 8 o'clock class?

[Ans. $n(\overline{A}) = 11$ from the diagram in Figure 3.6]

(b) How many of the students who like 8 o'clock class think that students should not be allowed to bring food to Classroom?

[Ans. $n(A \cap \overline{B}) = 10$ from the diagram in Figure 3.6]

(c) How many of the students who do not like 8 o'clock class think that students should be allowed to bring food to Classroom?

[Ans. $n(B \cap \overline{A}) = 6$ from the diagram in Figure 3.6]

(d) How many of the students who do not like 8 o'clock class think that students should not be allowed to bring food to Classroom?

[Ans. $n(\overline{A \cup B}) = n(U) - n(A \cup B) = 5$ from the diagram in Figure 3.6].

Example 3.1

1. In a survey, 100 students were asked the following 'Yes - No' questions:

(i).   Do you like Hamburgers?

(ii).   Do you like Pizza?

A total of 40 students answered 'Yes' to (i), a total of 80 students answered 'Yes' to (ii) and 25 students answered 'Yes' to both questions. Draw a Venn diagram to show the result of this survey and find each of the following:

    (a)   How many students either like Hamburger or Pizza?

    (b)   How many students like neither Hamburger nor Pizza?

    (c)   How many students like Pizza but not Hamburger?

    (d)   How many students like Hamburger but not Pizza?

Solution. Let   H be the set of students who answered 'Yes' to (i) and P be the set of students who answered 'Yes' to (ii). Then on the Venn diagram, first find the number that would go into H « P. The number of those who answered 'Yes' to both is given to be 25. So, $n(H « P) = 25$. Since $n(H) = 40$, the number that would be in the region inside H but outside H « P will be   40 - 25 = 15. Similarly, to find the number in P but outside H « P, calculate 80 - 25 = 55. The total number in H » P will be 15 + 25 + 55 = 95. So, 100 - 95 = 5 will be outside H » P and we will have the following Venn diagram:

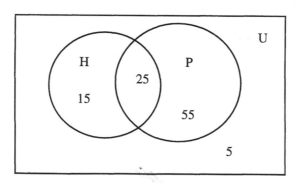

From the above work, the answers are:

(a) $n(H \cup P) = 15 + 25 + 55 = 95$

(b) $n(\overline{H \cup P}) = n(U) - n(H \cup P) = 100 - 95 = 5$

(c)   55

(d)   15.

Exercises 3.1.

1. In a survey, 30 students were asked the following 'Yes - No' questions:

   A. Do you drink coffee while doing homework?

   B. Do you watch TV while doing homework?

   A total of 15 students answered 'Yes' to A, a total of 18 students answered 'Yes' to B and 10 students answered 'Yes' to both questions. Draw a Venn diagram to show the result of this survey and find each of the following:

   (a) How many students either drink coffee or watch TV while doing homework?

   (b) How many students neither drink coffee nor watch TV while doing homework?

   (c) How many students watch TV but do not drink coffee while doing homework?

   (d) How many students drink coffee but do not watch TV while doing homework?

2. In a survey, 50 students were asked the following 'Yes - No' questions:

   A. Do you like Math?

   B. Do you like History?

   A total of 20 students answered 'Yes' to A, a total of 30 students answered 'Yes' to B and 5 students answered 'Yes' to both questions. Draw a Venn diagram to show the result of this survey and find each of the following:

   (a) How many students either like Math or History?

   (b) How many students neither like History nor Math?

   (c) How many students like Math but not History?

   (d) How many students like History but not Math?

3. In a survey of 100 students, it was found that 60 students drink, 50 students smoke, and 20 students both smoke and drink. Draw a Venn diagram to show the result of this survey and find each of the following:

   (a) How many students either drink or smoke?

   (b) How many students neither drink nor smoke?

4. In a survey of 50 students, it was found that 38 played soccer, 15 played football, and 7 played both football and soccer.

   (a) How many students play either soccer or football?

   (b) How many students play neither soccer nor football?

5.   Let A and B be two sets such that n(A) = 4, n(B) = 7, and n(A ∪ B) = 9. Find n(A ∩ B).

6.   Let A and B be two sets such that n(A ∪ B) = 20, n(A ∩ B) = 0, and n(A) = 6. Find n(B).

7.   Let A and B be two sets in a universal set U and such that n(A) = 12, n(B) = 13, n(U) = 25 and n(A ∪ B) = 19. Then find each of the following:

  (a) n($\overline{A}$) (b) n(A ∩ B) (c) n($\overline{U}$)

  (d) n(A ∩ $\overline{B}$) (e) n ($\overline{A ∪ B}$).

8.   Let A and B be two sets in a universal set U and such that n(A) = 15, n(B) = 11, n(U) = 30 and n(A ∪ B) = 21. Then find each of the following:

  (a) n($\overline{A}$) (b) n(A ∩ B) (c) n(∅)

  (d) n(A ∩ $\overline{B}$) (e) n ($\overline{A ∪ B}$).

9.   A deck of cards has 4 suits, namely clubs, diamonds, hearts and spades. Each suit has 4 honor cards which are the ace, king, queen and jack, and 9 numeric cards numbered from 2 to 10. Figure 3.7 shows the 52 cards in a deck.

Suit	Symbol (color)	Honor cards	Ordinary cards
Clubs	♣(black)	♣A, ♣K, ♣Q, ♣J	♣2, ♣3, ♣4, ♣5, ♣6, ♣7, ♣8, ♣9, ♣10
Diamond	♢(red)	♢A, ♢K, ♢Q, ♢J	♢2, ♢3, ♢4, ♢5, ♢6, ♢7, ♢8, ♢9, ♢10
Hearts	♡(red)	♡A, ♡K, ♡Q, ♡J	♡2, ♡3, ♡4, ♡5,♡6, ♡7, ♡8, ♡9, ♡10
Spade	♠(black)	♠A, ♠K, ♠Q, ♠J	♠2, ♠3, ♠4, ♠5, ♠6, ♠7, ♠8, ♠9, ♠10

Figure 3.7. Cards in a deck

Let   U be the set of cards,   B = The set of black cards in a deck,

  P = The set of face cards

    = {♣K, ♣Q, ♣J, ♢K, ♢Q, ♢J, ♡K, ♡Q, ♡J, ♠K, ♠Q, ♠J}

Find each of the following:

  (a) n(B) (b) n(P) (c) n($\overline{P}$)

  (d) n(B ∩ P) (e) n(B ∪ P) (f) n($\overline{B ∪ P}$)

10. Let   U be the set of cards in a deck (see Figure 3.7) and

    D = The set of diamonds in a deck

    V = The set of honor cards in a deck.

    Find each of the following:

    (a) n(V)   (b) n($\overline{V}$)   (c) n(V ∩ D)

    (d) n(V ∪ D)   (e) n($\overline{V \cup D}$)

11. Let   U be the set of cards in a deck (see Figure 3.7) and

    H = The set of hearts in a deck

    P = The set of face cards.

    Find each of the following:

    (a) n(H)   (b) n($\overline{H}$)   (c) n(H ∩ P)

    (d) n(H ∪ P)   (e) n($\overline{H \cup P}$).

12. Let   U be the set of cards in a deck (see Figure 3.7) and

    H = The set of hearts in a deck

    R = The set of red cards

    Find each of the following:

    (a) n(R)   (b) n($\overline{R}$)   (c) n(H ∩ R)

    (d) n(H ∪ R)   (e) n($\overline{H \cup R}$).

The idea of sets and counting can be applied to situations other than surveys. We need to look into some of these other situations. Therefore, we give a formal discussion of set theory.

2. SETS AND ELEMENTS

In mathematics, one learns to organize both ideas and expressions. The language and notation of set theory provides a good tool to do this. The basic idea is to put things like numbers, points, people, etc., into boxes called 'sets' as done in section 1. Thus, we may put the numbers 1, 2 and 3 in a box called a set whose elements are 1, 2 and 3. We write this as:

$$\{1, 2, 3\} \text{ (the set whose elements are 1, 2 and 3)},$$

where { } stands for the expression 'the set'. We say that 1 belongs to {1, 2, 3} but 4 does not belong to {1, 2, 3}. The grouping symbol { } is used to denote a set, and the elements of the set, separated by commas, are listed inside { }. In set theory, it is therefore advisable that one does not use the braces { } as a grouping symbol.

A set is a single entity containing certain members or elements. A set is *well-defined* if there is no ambiguity as to what belongs to the set. One good thing about sets is their universality. Virtually anything can be made into sets. We can create sets of numbers, sets of letters, sets of people, sets of objects, and even sets of sets.

Examples 3.2

1. Sets with one element each: {0}, {1}, {2}, {3}, {10}, {-17}, {-12}{Michael}.

2. Sets with two elements each: {0, 1}, {1, 2}, {3, -3}, {13, 10}, {M, N}.

3. Sets with three elements each: {0, 1, -1}, {1, 2, 3}, {13, 10, 0}, {1, -17, -12}.

4. Sets with four elements each: {0, 1, -1, 2}, {1, 2, 3, 4}, {0, 1, -17, -12}.

5. Sets with five elements each: {0, 1, -1, 2, -2}, {1, 2, 3, 4, 5}, {a, e, i, o, u}.

6. {1, 2, {0, 3}}. This is a set with three elements. Two of the elements are numbers and the other element is a set.

7. The set of letters in the English alphabet.

The Empty Set

We can also have a set with no elements. This is called the *empty set* and it is denoted by ∅. As will be seen later, the empty set plays a very important role in mathematical formalism. It should be noted that ∅ ≠ {∅} since {∅} is a set which has the empty set as an element whereas ∅ is a set which has no element in it.

Naming Sets

As mentioned earlier, it is convenient to give names to sets. When naming sets, we use capital letters to name a set and use lower case letters to name the elements. For example, we may name the set {a, b, c} as A. We write let A = {a, b, c}. In this case, A itself stands for a set and it is not necessary to enclose it inside the braces. As a matter of fact, {A} will mean a set with one element, namely, the set A. Since the phrases 'belongs to' and 'does not belong to' are used frequently, we have the following symbols for these phrases.

$\in$: belongs to

$\notin$: does not belong to

Sets and elements are distinguished by the following rule.

> RULE-1. A set cannot be an element of itself.

In other words, if S is a set, then $S \notin S$.

DEFINITION 3.1. Two sets A and B are equal if every element of A is also an element of B and if every element of B is also an element of A. If A and B are equal then we write A = B. Therefore, $A \neq B$ means that set A is not equal to the set B.

Examples 3.3

1. Let A = {0, 1, -1, 2}. Then $0 \in A, 1 \in A$ but $5 \notin A$.

2. Let B = {1, 2, 3, 4, 5, 6}. Then $1 \in B, 5 \in B$ but $7 \notin B$

3. $1 \in \{1, 2, 3\}$ but $\{1\} \notin \{1, 2, 3\}$ because {1} denotes a set which is not an element of the set indicated by {1, 2, 3}.

4. We cannot write $\{1, 2\} \in \{1, 2\}$ because of Rule-1.

5. If A is a set, then $A \in \{A\}$ but $A \notin A$.

6. {1, 2, 3} = {3, 1, 2}

7. If A = {1, 2, 3,. . . , 10} and B = {a, e, i, o, u}, then $A \neq B$.

Exercises 3.2

1. Give an example of each of the following:
    (a) A set with one element.
    (b) A set with four elements.

2. Give an example of each of the following:
    (a) A set with three elements.
    (b) A set with no elements.

3. State the difference between   (a) 1 and $\{1\}$ (b) $\varnothing$ and $\{\varnothing\}$

4. State the difference between   (a) 1, 2 and $\{1, 2\}$ (b) 0 and $\varnothing$

5. Let A = $\{1, 2, 3, \{1\}, \{1, 2\}\}$.
    (a) What are the elements of A?
    (b) Is the statement $\{1, 2\} \in A$ true?
    (c) Is the statement $\{2\} \in A$ true?
    (d) Can one set be the element of another set?
    (e) Is the statement $\{3\} \in A$ true?
    (f) Is the statement $3 \in A$ true?

6. Let A = $\{0, 1, \varnothing\}$.
    (a) What are the elements of A?
    (b) Is the statement $\{\varnothing\} \in A$ true?
    (c) Is the statement $\{1\} \in A$ true?
    (d) Can the empty set be an element of another set?
    (e) Is the statement $\varnothing \in A$ true?
    (f) Is the statement $0 \in A$ true?

Universal sets

As done in section 1, we represent a set A by the region inside a closed curve. A point a inside the region A will mean $a \in A$. We regard every set as contained in a universal set U. A specific set U is a universal set for a given situation if U contains all set theoretic elements of the situation under consideration. For example, in our survey example the set of students in the class was the universal set. A universal set may vary from situation to situation. One use of the universal set is in describing the complement of a set, namely, the set consisting of all the elements not in a given set.

The set builder notation

To describe $\overline{A}$, it is not enough to say that it consists of all elements $x \notin A$ because anything or anybody that do not belong A would be in $\overline{A}$. However, this problem is avoided by using a universal set U. Thus we say $\overline{A}$ (the *complement of the set* A) consists of those elements of U which are not in A [See Figure 3.1]. This can be expressed in a compact form by using the set builder notation.

$$\overline{A} = \{x \in U \mid x \notin A\}. \quad (\text{the set of all } x \in U \text{ such that } x \notin A).$$

In the *set builder notation* { | }, a vertical line (or colon) is used to separate the expression inside the set notation into two parts. The first part tells where each element comes from or what each element is represented by. The second part gives a characterizing property of each element. The line stands for 'such that'. Thus,

$$\{ x \mid p \} \quad (\text{the set of all x such that p}),$$

where p is usually a statement characterizing the elements of the set. The use of the letters x and p in this notation is arbitrary and other letters may be used in their places.

Note. $\qquad \overline{U} = \emptyset$

$$\overline{\emptyset} = U.$$

Sets do not involve operations like addition or multiplication. Instead, when dealing with sets, we can take what are called union and intersection.

The Intersection of sets

If we take the set of elements that belong to both of two sets, then we get the intersection of the two sets. The intersection of two sets is denoted by '∩'. If A and B are two sets, then

$$A \cap B \quad (\text{A intersection B})$$

is the set of all elements that are in A as well as B.

For example, if A = {0, 1, -1, 2} and B = {1, 2, 3, 4, 5, 6}, then

$$A \cap B = \{1, 2\}.$$

The union of two sets

If we put the elements of two sets together, then we get a new set called the *union* of the two sets. The symbol for union is '∪'. If A and B are two sets, then

$$A \cup B \quad (A \text{ union } B)$$

is the set consisting of all the elements of A together with the elements of B.

For example, if A = {0, 1, -1, 2} and B = {1, 2, 3, 4, 5, 6}, then
A ∪ B = {0, 1, -1, 2, 3, 4, 5, 6}.

Note. If the same element appears in both sets, then we write it only once in the union.

For each element of A or B, one should ask the questions "Is the element in A?", "Is the element in B?" If the answer is 'yes' to any one or both of these questions then the element is in A ∪ B. Given the diagrams for sets A and B, the region A ∪ B will be represented by the region inside A together with the region corresponding to B as shown in Figure 3.4 (figure on right). Below, we give some more examples of union and intersection of sets. For each element of A or B, one should ask the questions: Is the element in A? Is the element in B? If the answer is 'yes' to both of these questions then the element is in A ∩ B. For a Venn diagram of the intersection A ∩ B, see Figure 3.4. The set builder notation may be used to give formal definitions of union and interesection.

DEFINITION 3.2. The *union* A ∪ B of two sets A and B is defined by

$$A \cup B = \{x \mid x \in A \text{ or } x \in B\}.$$

DEFINITION 3.3. The intersection A ∩ B of two sets A and B is defined by

$$A \cap B = \{x \mid x \in A \text{ and } x \in B\}.$$

Disjoint sets. Two sets are disjoint if their intersection is empty.

For example, A = {1, 2, 3} and B = {a, e, i, o, u} are disjoint sets. If A and B are disjoint then their diagrams are two non-intersecting closed curves (Figure 3.8).

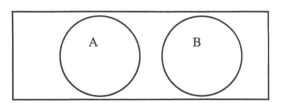

Figure 3.8. Disjoint sets

Examples 3.4

1. Let U = {1, 2, 3,. . . , 10, 11, 12}, A = {1, 2, 3, 4, 5, 6} and

   B = {2, 4, 6, 8, 10}. Then [see Figure 3.9]

   A ∩ B = {2, 4, 6},   A ∪ B = {1, 2, 3, 4, 5, 6, 8, 10}

   $\overline{A}$ = {7, 8, 9, 10}, $\overline{B}$ = {1,3,5,7,9},

   A ∩ $\overline{B}$ = {1, 3, 5}   and   B ∩ $\overline{A}$ = {8, 10}.

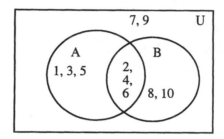

Figure 3.9

2. Let U = {1, 2, 3,. . . , 10},
   A = {x ∈ U I x is even}   B = {x ∈ U I x is a multiple of 3}.

   A = {2, 4, 6, 8, 10},   B = {3, 6, 9},

   $\overline{A}$ = {1, 3, 5, 7, 9}, $\overline{B}$ = {1, 2, 4, 5, 7, 8, 10},

   A ∩ B = {6}   A ∪ B = {2, 3, 4, 6, 8, 9, 10}.

Exercises 3.3

1. Let U = {1, 2, 3,. . . , 20}, A = {1, 2, 3,. . . , 10}, and B = {2, 4, 6,. . . , 20}. Find
   each of the following sets:

   (a) $\overline{A}$ (b) $\overline{B}$ (c) A ∩ B   (d) A ∩ $\overline{B}$
   (e) B ∩ $\overline{A}$ (f) $\overline{A \cup B}$.

2. Let U = {1, 2, 3,. . . ,15}, A = {x ∈ U | x is a multiple of 4} and B = {3, 8}. Find each of the following sets:

     (a) $\overline{A}$ (b) $\overline{B}$ (c) A ∩ B   (d) A ∩ $\overline{B}$

     (e) B ∩ $\overline{A}$ (f) A ∪ B .

3. Let U = {0, 1, -1, 2, -2, 3, -3, 4, -4, 5, -5, . . . , 10, -10},

     A = {x ∈ U | x is positive} and B = {x ∈ U | x is a multiple of 3}.

     A = { 1, 2, 3, 4, 5, 6, 7, 8, 9, 10 }   *B = {-3 3, -6,6, -9,9}*

     Find each of the following sets:

     (a) $\overline{A}$ (b) $\overline{B}$ (c) A ∩ B   (d) A ∩ $\overline{B}$

     (e) B ∩ $\overline{A}$ (f) A ∪ B .

4. Let U = {0, 1, -1, 2, -2, 3, -3, 4, -4, 5, -5, . . . , 10, -10},

     A = {x ∈ U | x is negative} and B = {x ∈ U | x is a multiple of 4}.

     Find each of the following sets:

     (a) $\overline{A}$ (b) $\overline{B}$ (c) A ∩ B   (d) A ∩ $\overline{B}$

     (e) B ∩ $\overline{A}$ (f) A ∪ B .

5. Let U = the set of letters in the English alphabet,

     A = the set of letters in the sentence 'Mathematics is fun',

     B = the set of consonants.

     Find each of the following:

     (a) $\overline{A}$ (b) $\overline{B}$ (c) A ∩ $\overline{B}$ (d) B ∩ A.

6. Let U = the set of letters in the English alphabet,

     A = the set of letters in the phrase, 'I want to hold your hand"

     B = the set of consonants.

     Find each of the following:

     (a) $\overline{A}$ (b) $\overline{B}$ (c) A ∩ $\overline{B}$ (d) B ∩ A.

7. Let U = {1, 2, 3,. . . ,10}, A = {x ∈ U| x is a multiple of 3}, B = {x ∈ U| x is odd}.

     Find each of the following sets: *{ 3, 6, 9 }*   *{1, 3, 5, 7, 9}*

     (a) {x ∈ B | x ∈ A}   (b) {x ∈ B | x ∉ A}   (c) {x ∈ A | x ∉ B}

     (d) {x ∈ A | x ∈ B}   (e) {x ∈ U | x ∉ A and x ∉ B}

8. Let U = {1, 2, 3,. . . , 20},   A = {x ∈ U| x is a multiple of 5} and
    B = {x ∈ U| x is even}. Find each of the following sets:

    (a) {x ∈ B | x ∈ A}   (b) {x ∈ B | x ∉ A}

    (c) {x ∈ A | x ∉ B}   (d) {x ∈ U | x ∉ A and x ∉ B}

9. Let U = {1, 2, 3,. . . ,100}. Express each of the following sets in set builder notation:
    (a) {4, 8, 12,. . .96, 100 }   (b) {1, 2, 3,. . . 10}
    (c) {5, 10, 15, 20, . . . , 100}   (d) {10, 20, 30,. . .,100}
    (e) {1, 4, 9, 16,. . ., 100}

10. Let U = {1, 2, 3,. . . , 50}. Express each of the following sets in set builder notation:
    (a) {4, 8, 12,. . ., 48}   (b) {1, 3, 5, . . . , 49}
    (c) {5, 10, 15, 20, . . . , 50}   (d) {10, 20, 30,. . .,50}
    (e) {1, 4, 9, 16,. . ., 49}   (f) { 2, 3, 5,. . ., 47}

11. Let U = {1, 2, 3,. . . , 50}. Describe each of the following sets by giving at least 3
    elements:

    (a) {x ∈ U | x is prime}   (b) {x ∈ U | x is a multiple of 5
    (c) {x ∈ U | x is a multiple of 6} (d) {x ∈ U | x is a perfect square}

12. Let U = {1, 2, 3,. . . ,100}. Give at least 3 elements of each of the following:
    (a) {x ∈ U | x is a perfect square} (b) {x ∈ U | x is a multiple of 7}
    (c) {x ∈ U | x is a multiple of 15} (d) {x ∈ U | x > 90}.

13. Let S and T be two sets. Define S ∪ T using set builder notation.

14. Let S and T be two sets. Define S ∩ T using set builder notation.

15. Which of the following pairs of sets are disjoint?
    (a) {a, b}, {a, b, c}   (b) {1, 2}, {-1, -2}   (c) {0, 1}, {∅}
    (d) {1, 2}, {a, e, i}.

3. SUBSETS

In case of a deck of cards, if H = the set of hearts in the deck, and R = the set of red cards in the deck, then we see that every element of H is an element of R since every heart is red. In the set diagram of this situation the region for H will be completely inside the region for R. However, we cannot write H ∈ R because the symbol '∈' can be used only for elements and not for a set within a set and H is not an elemnt of R. To describe this situation, we use the term subset and the notation '⊂'.

If H ⊂R, then H and R may be represented by the diagram shown in Figure 3.10.

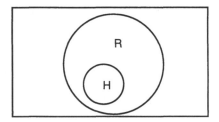

Figure 3.10.   H ⊂ R

DEFINITION 3.4. A set A is a *subset* of a set B if every element of A is also an element of B. In this case, we write

$$A \subset B \text{ (A is a subset of B)}$$

We have the following rules:

> RULE - 2. *Every set is a subset of itself.*
>
>   That is, if A is any set, then A ⊂ A.

> RULE - 3. *The empty set is a subset of every set.*
>
>   That is, for any set A, ∅ ⊂ A.

DEFINITION 3.5. Let A be a non-empty set. Then ∅ and A are *improper subsets* of A. Any subset of A other than ∅ and A is a *proper subset* of A.

Examples 3.5

1. H $\subset$ R.

2. P $\subset$ V, where P is the set of face cards in a deck and V is the set of honor cards in the deck.

3. Let A = {x | x is a son of Mr. X} and B = {x | x is a son or daughter of Mr. X}. Then A may be an improper subset of B since Mr. X may not have a son, in which case, A = $\varnothing$. Or, Mr. X may not have a daughter in which case A = B.

4. $\varnothing \subset$ {1,2,3}.

5. Let A = {a, b, c}. Then we may find all the subsets of A as follows:

    Solution. The two improper subsets are $\varnothing$ and {a,b,c}. Any other subset of A must have either one or two elements. Therefore, we get the following subsets of A are $\varnothing$, {a}, {b}, {c}, {a, b}, {b, c}, {c, a}, and {a, b, c}.

THEOREM 3.1. *If A $\subset$ B and B $\subset$ A then A = B. Conversely, if A = B then A $\subset$ B and B $\subset$ A.*

Proof. If A $\subset$ B and B $\subset$ A then every element of A is in B and every element of B is in A so that A = B. Similarly, if A = B then every element of A is also in B (A $\subset$ B) and every element of B is in A (B $\subset$ A). ∎

Exercises 3.4

1. Let A = {1, 2}. Find all the subsets of A.

2. Let A = {1}. Find all the subsets of A.

3. (a) How many proper subsets does A = {1, 2, 3} have?

    (b) How many improper subsets does a non-empty set have?

    (c) How many proper subsets can the empty set have?

    (d) How many improper subsets does the empty set have?

    (e) Is the set of rational numbers a subset of the irrational numbers?

4.   (a) How many proper subsets does A = {1, 2, 3, 4} have?

(b) How many improper subsets does A = {1, 2, 3, 4} have?

(c) Give the proper subsets of A = {1, 2, 3, 4} that have one element each?

(d) Give the proper subsets of A = {1, 2, 3, 4} that have two elements each?

(e) Is the set {1, 2, 3} a proper subsets of A = {1, 2, 3, 4}?

5.   For each of the following pairs of sets write a true statement using $\subset$ and the given sets.

(a)   B = The set of black cards in a deck, S = The set of spades in a deck

(b)   V = the set of honor cards in a deck of cards,

K = The set of kings in a deck of cards,

(c) $\varnothing$, A = {a, b, c}

(d)   A = {1, 2, 3, 4}, B = {2, 4}.

6.   For each of the following pairs of sets write a true statement using $\subset$ and the given sets.

(a)   R = The set of red cards in a deck, S = The set of hearts in a deck

(b)   F = the set of honor cards in a deck of cards,

Q = The set of queens in a deck of cards,

(c) $\varnothing$, {$\varnothing$}

7.   Give all the subsets of each of the following:

(a) $\varnothing$ (b) {a}   (c) {a, b}   (d) {a, b, c}

8.   Give all the subsets of each of the following:

(a) {1}   (b) {1, 2}   (c) {1, 2, 3}

9.   Let A = {1, 2, 3, 4}. Find the subsets of A that are not subsets of {1, 2, 3}.

10.   Give all the subsets of {a, b, c, d} that are not subsets of {a, b, c}.

4. ONE-TO-ONE CORRESPONDENCE

DEFINITION 3.6. Let A and B be two sets. We say that A and B are in a *one-to-one correspondence* (1-1 correspondence) if we can assign to each element of A one partner, an element of B, such that every element of B is a partner of one and only one element of A. In this case, we write

$$A \leftrightarrow B \text{ (A and B are in one-to-one correspondence)}.$$

If $b \in B$ is the partner assigned to $a \in A$, then b corresponds to a and we write $a \leftrightarrow b$. Thus, A = {a, b, c} and B ={1, 2, 3} are in one-to-one correspondence because $a \leftrightarrow 1$, $b \leftrightarrow 2$, $c \leftrightarrow 3$.

There is no way that one can set up a one-to-one correspondence between A = {a, b, c} and C ={1, 2, 3, 4}.

DEFINITION 3.7. A set S is *finite* if it is empty or if there exists a natural number p such that the set S is in one-to-one correspondence with {1, 2, 3. . . , p}. In the latter case, S has p elements and we write

$$n(S) = (\text{number of elements in S}) = p.$$

We take $\quad n(\varnothing) = 0.$

Counting the number of elements in a set means finding a natural number p such that the set {1, 2, 3,. . . , p} is in one-to-one correspondence with the given set.

DEFINITION 3.8. A set is *infinite* if it is not finite.

The simplest infinite set is the set of *natural numbers*. It is denoted by N. So,
$$N = \{1, 2, 3, 4, 5, \ldots\}.$$

Another important infinite set is the *set of integers* Z. It includes the natural numbers, zero and the negative whole numbers. So,
$$Z = \{0, 1, -1, 2, -2, 3, -3, 4, -4, 5, -5, \ldots\}.$$

Exercises 3.5

1. Which of the following sets are in one-to-one correspondence?

A = {1, 2, 3, 4}	B = {a, b, c}	C = {x, y, z, t}
D = {$\frac{1}{2}$, $\frac{1}{3}$, $\frac{1}{4}$}	E = {3, 6, 9}	F = {0}
G = $\varnothing$	H = {1, 2, 3,. . . }	I = {2, 4, 6,. . . }.

2.   Which of the following sets are in one-to-one correspondence?

A = {a, e, i, o, u}   B = {1, 2}   C = {2, 2^2, 2^3,. . . }

D = {$\frac{1}{2}$}   E = {1, 2, 3, 4, 5}   F = {0}

G = {N, Q}   H = {1, 2, 3,. . . }   I = {1, $\frac{1}{2}$, $\frac{1}{3}$,. . . }.

3.   In each of the following determine if the given set is finite or infinite. If the set is finite find the number of elements in it.

(a) {1, 2, 3,. . . , 15}   (b) N = {1, 2, 3,. . . }

(c) {2, 4, 6,. . . }   (d) {x ∈ N | 1 < x < 10}

(e) {x ∈ N | x is a multiple of 3}   (f) {N}

(g) The set of letters in the English alphabet.

4.   In each of the following determine if the given set is finite or infinite. If the set is finite find the number of elements in it.

(a) {-1, -2, -3,. . . , -15}   (b) {x ∈ Z | x is a multiple of 3}

(c) The set of even integers   (d) {x ∈ Z | -1 < x < 10}

(e) {x ∈ Z | -1 ≤ x ≤ 10}   (f) {x ∈ Z | x is a multiple of 4}

5.   Let N = {1, 2, 3,. . . }. In each of the following determine if the given set is finite or infinite. If the set is finite find the number of elements in it.

(a) {x ∈ N | x < 10}   (b) {x ∈ N | x > 10}

(c) {x ∈ N | x is even}   (d) {x ∈ N | 5 < x < 10}

(e) {x ∈ N | 1 ≤ x ≤ 10}   (f) {x ∈ N | x > 15}

(g) {x ∈ N | 3 < x < 10}   (h) {x ∈ N | 5 ≤ x ≤ 10}

6.   Let Z be the set of integers. In each of the following determine if the given set is finite or infinite. If the set is finite find the number of elements in it.

(a) {x ∈ Z | x < 10}   (b) {x ∈ Z | x > 10}

(c) {x ∈ Z | x is even}   (d) {x ∈ Z | -5 < x < 10}

(e) {x ∈ Z | 0 ≤ x ≤ 10}   (f) {x ∈ Z | x > 15}

(g) {x ∈ Z | -3 < x < 10}   (h) {x ∈ Z | -5 ≤ x ≤ 10}

5. COUNTING WITH MORE THAN TWO SETS

So far, we have been working with the union and intersection of only two sets. We can take the union and intersection of more than two sets as well. To do so, we will need to use parentheses. Remember that we always work inside parentheses first.

Examples 3.6

1. Let $A = \{0, 1, -1, 2, -2, 3, -3\}$, $B = \{1, 3, 5, 7\}$ and $C = \{2, 4, 6\}$.
    Find each of the following sets:

    (a) $(A \cup B) \cap C$ (b) $A \cap (B \cup C)$

    (c) $(A \cap C) \cup B$ (d) $A \cap B \cap C$.

    Solutions.   (a) $\{0, 1, -1, 2, -2, 3, -3, 5, 7\} \cap \{2, 4, 6\} = \{2\}$

    (b) $\{0, 1, -1, 2, -2, 3, -3\} \cap \{1, 2, 3, 4, 5, 6, 7\} = \{1, 2, 3\}$

    (c) $\{2\} \cup \{1, 3, 5, 7\} = \{1, 2, 3, 5, 7\}$

    (d) $\varnothing$.

2. In a survey of 40 persons it was found that 5 persons smoked but were not alcoholics, 22 persons had cancer, 16 persons were alcoholics and had cancer, 12 were alcoholics but did not have cancer, 17 persons smoked and were alcoholics, 15 persons smoked, were alcoholics, and had cancer. If there was only one person who smoked but was neither an alcoholic nor had cancer, then

    (a)   How many persons did not smoke?

    (b)   How many persons smoked and had cancer?

    (c)   How many persons did not smoke and did not get cancer?

    (d)   How many alcoholics did not smoke and did not get cancer?

    Solution. Let $S = \{$persons who smoked$\}$, $A = \{$alcoholics$\}$,
              $C = \{$persons who had cancer$\}$ and $U = $ Universal set. Then

    $n(U) = 40,$ $n(S \cap \overline{A}) = 5,$ $n(C) = 22,$

    $n(A \cap C) = 16,$ $n(A \cap \overline{C}) = 12,$ $n(S \cap A) = 17,$

    $n(S \cap A \cap C) = 15,$ and $n(S \cap (\overline{C \cup A})) = 1.$

Inserting these numbers into the appropriate regions of the Venn diagram, we get

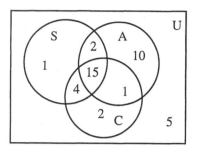

Figure 3.11

(a) $n(\overline{S}) = n(U) - n(S) = 40 - 22 = 18,$

(b) $n(S \cap C) = 19,$

(c) $n(\overline{S} \cap \overline{C}) = n(\overline{S \cup C}) = 10 + 5 = 15,$

(d) $n(A - (S \cup C)) = 10.$

3.   To see which topics he should he review before the test, a teacher asked the students to complete a Yes-No questionnaire. The first three questions were:

Do you know how to find
I.   the mean deviation of a data?   Yes   No
II.   the standard deviation of a data?   Yes   No
III.   the mean of a grouped data?   Yes   No.

The table in Figure 3.12 gives the answers by 27 students. Display the data on a Venn diagram and answer each of the following:
(a) In which topic the class is least prepared?
(b) How many students knew how to find mean deviation but not standard deviation?
(c) How many students were prepared for the test?
(d) How many students did not know how to find the mean of a grouped data?

Students	Question 1	Question 2	Quesion 3
1.	Yes	Yes	No
2.	Yes	Yes	Yes
3.	Yes	Yes	Yes
4.	No	No	Yes
5.	No	No	No
6.	Yes	No	Yes
7.	Yes	Yes	Yes
8.	Yes	No	No
9.	Yes	No	No
10.	Yes	No	No
11.	No	No	No
12.	Yes	Yes	Yes
13.	Yes	Yes	Yes
14.	Yes	Yes	Yes
15.	Yes	No	Yes
16.	Yes	No	No
17.	No	Yes	Yes
18.	No	Yes	No
19.	Yes	Yes	Yes
20.	Yes	Yes	Yes
21.	Yes	No	Yes
22.	Yes	No	Yes
23.	Yes	No	No
24.	Yes	Yes	No
25.	No	No	No
26.	No	No	Yes
27.	Yes	No	Yes

Figure 3.12

(a) Standard deviation

(b) 10

(c) 8

(d) 11

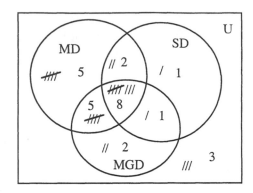

Exercises 3.6

1. Let A = {2, -2, 4, -4}, B = {0, 2, -2}, C = {2, 4, 6}.

   Find each of the following sets:

       (a) (A ∪ B) ∩ C   (b) A ∩ (B ∪ C)

       (c) (A ∩ C) ∪ B   (d) A ∩ B ∩ C.

2. Let A = {0, 1, -1, 2, -2, 3, -3},   B = {1, 3, 5, 7},   C = {1, 3, -3}. Find
   each of the following sets:

       (a) A ∪ (B ∩ C)   (b) (A ∩ B) ∪ C

       (c) A ∩ (C ∪ B)   (d) A ∩ B ∩ C.

3. In a survey of 50 joggers, it was found that 30 were jogging for fun, 32 were
   trying to lose weight, 18 were trying to lose weight and were having fun, 14 were
   staying away from work and were having fun, 17 were trying to lose weight and
   were trying to stay away from their work, 10 were trying to stay away from their
   work, were trying to lose weight, and were having fun. If there was only one
   jogger who was trying to stay away from his work but was neither trying to lose
   weight nor having fun, then draw a Venn diagram summarizing the data and find
   each of the following:

   (a) How many joggers were neither trying to stay away from their work, nor
   trying to lose weight, nor having fun?

   (b) How many were trying to stay away from their work?

4. In a survey of 100 students, the numbers of students studying Math, Physics, and
   Chemistry were found to be Math 45, Phy 40, Chem 25, Math and Phy 20, Phy

and Chem 12, Chem and Math 10, all three 5. Draw a Venn diagram and answer each of the following:

(a) How many students were studying none of the subjects, Math, Phy, or Chem?

(b) How many students were studying Math only?

(c) How many students were studying either Math or Phy?

(d) How many students were studying neither Chem nor Phy?

5. In a survey, 150 students were asked the following Yes/No questions:

Questionnaire table

I.	Do you approve of the Math requirement ?	Yes	No
II.	Do you like Math ?	Yes	No
III.	Do you like to gamble ?	Yes	No

It was found that 50 answered 'Yes' to question I, 75 to question II and 70 to question III; 25 answered 'Yes' to the first two questions, 15 answered 'Yes' to the first and the third question, 30 answered 'Yes' to the last two questions. If there were 10 students who answered 'Yes' to all three questions, then

(a) How many students did not approve of the Math requirement?

(b) How many students who like Math and also like to gamble approved of the math requirement?

(c) How many students who like Math and do not like to gamble approved of the requirement?

(d) How many students answered 'No' to each of the three questions?

6. You are required to make a survey of the scores in test 1 of this course to determine each of the following:

(a) How many students who received an 'A' attend classes regularly?

(b) How many students who do not hand in homework received an 'A'?

(c) How many students who attend classes regularly but do not hand in home work regularly received an 'A'?

(d) How many students who do not attend classes regularly and do not hand in home work regularly received an 'A'?

State how you would conduct the survey to answer the above questions.

CHAPTER TEST 3

1.   Let U = {1, 2, 3, 4, 5, 6, 7, 8, 9, 10},

      A = {x ∈ U | x is odd},   B = {4, 6, 8},   C = {1, 2,3}.

   Find each of the following sets:

   (a) A ∩ C   (b) (A ∪ B) ∩ $\overline{A}$.

2.   State the difference between

   (a) ∅ and {∅}   (b) 3, 4 and {3, 4}   (c) A and $\overline{A}$

   (d) A and n(A)   (e) a finite set and an infinite set.

3.   Let U = {1, 2, 3,. . . ,100}. Express each of the following sets in the set builder
      notation:

   (a) {3, 6, 9, . . . , 99 }   (b) {2, 4, 6, . . . , 100}

   (c) {2, 4, 8, 16, . . . , 64}   (d) { 2, 3, 5, . . ., 97}

4.   Let U = {0, 1, -1, 2, -2, 3, -3, 4, -4, 5, -5, . . . , 10, -10}

      A = {x ∈ U | x is positive}

      B = {x ∈ U | x is a multiple of 4}.

   Find each of the following sets:

   (a) $\overline{A}$ (b) $\overline{B}$ (c) A ∩B   (d) A ∩ $\overline{B}$

   (e) B ∩ $\overline{A}$ (f) $\overline{A ∪ B}$.

5.   Let A = {a, b, c}. Give
      (a) all the proper subsets of A
      (b) all the improper subsets of A.

6.   To study student behavior, a defective Coke machine that took coins but did not
      give any Coke was placed in the student union and a survey was made of 200
      students who used the machine. It was found that 50 students kicked the machine
      and 100 swore at the machine. If there were 10 students who kicked and swore at
      the machine, give a Venn diagram summarizing the data and find:
      (a) How many students neither kicked nor swore at the machine?
      (b) How many students kicked but did not swear at the machine?

7.   Let   U be the set of cards and

        H = The set of hearts in a deck

        P = The set of face cards

Find each of the following:

(a) n(H)   (b) n(P)   (c) n($\overline{P}$)

(d) n(H ∩ P)   (e) n(H ∪ P)   (f) n($\overline{H \cup P}$)

8.   (a) Determine which of the following sets are in one-to-one correspondence.

    A = {1,2,3},   B = ∅,   C = {s, i, x},   D = {0},

    E = {1, 2},   F = {∅},   G = {δ, σ},

    H = {3, 6, 9,. . . }   and   N = the set of natural numbers.

(b) Determine which of the following sets are infinite.

    A = {2, 4, 8, 10},   B = {2, 4, 6, 8, 9, . . . }

    C = {1, 2, 3,. . . }   D = {1, 2, 3,. . . , 9},

    E = The set of letters in the English alphabet.

9.   Which of the following statements are false? For each false statement give a reason why it is false.

(a) 2 ⊂ {1, 2, 3}   (b)   1 ∈ {1, 3}   (c) $\overline{U}$ = ∅

10.   In a survey of 100 students, it was found that 40 were studying Math, 29 were studying Physics, 32 were studying Chemistry. There were 12 studying Math and Phy, 11 studying Physics and Chemistry, 10 studying Chemistry and Math. Only 4 of the students were studying all three subjects. Draw a Venn diagram showing the results of the survey and answer each of the following:

(a)   How many students were studying none of the subjects, Math, Physics, or Chemistry?

(b)   How many students were studying Math only?

(c)   How many students were studying either Math or Phy?

(d)   How many students were studying neither Chem nor Phy?

Chapter 4

COUNTING TECHNIQUES

1. TREE DIAGRAMS

Experiments and Sample spaces

The basic counting techniques are useful tools in probability theory, statistics, analysis of scientific data, and for many simple situations in life. Yet, one can master these techniques with the knowledge of elementary arithmetic. Because of their wide application in everyday life, the counting techniques are sources of fun and enjoyment. Our main goal is to find ways of counting the number of possible outcomes when a certain experiment is performed or when an event takes place. For example, we may want to solve decision problems like the following:

1. A basketball team of 5 players is to be formed from 8 possible players. How many ways can this be done?
2. A coin is tossed once and then a die is rolled once. How many different outcomes are possible?
3. A state projects that in the future, the number of cars to be registered in any given year will not exceed 500,000. What kind of system should the state select to number the license plates?
4. A soccer league has 10 teams. How many different games will there be if each team plays every team in the league exactly once?
5. Seven people are to be lined up for a photograph. How many different ways can this be done?

Each of the above describes an experiment with a number of outcomes that is to be determined. We accept the word *experiment* as a concept. We shall use the term 'experiment' in a broad sense to include anything whose outcomes may be determined. Here we will be concerned with experiments that have a finite number of outcomes. Thus, tossing a coin, rolling a die, drawing a ball from an urn, and drawing cards from a deck are all examples of experiments. With this understanding of the word experiment, we can define the following.

DEFINITION. The *sample space* of an experiment is the set of all possible outcomes of the experiment.

For any experiment under consideration, we take the set of all possible outcomes as the corresponding universal set and usually denote it by U.

Examples 4.1

1. If a coin is tossed once then the sample space U = {H,T}.

2. If a letter is chosen from the word 'mathematics', then the sample space is
    $$U = \{m, a, t, h, e, i, c, s\}.$$

3. If a die is rolled once then the sample space U = {1, 2, 3, 4, 5, 6}.

A finite sequence of two or more experiments is also an experiment. For example, if we toss a coin and then roll a die (see 1 of Examples 4.2). First, we shall consider some ways of listing the set of all possible outcomes of an experiment and then discuss various techniques of counting them.

A very useful method of listing all possible outcomes of an experiment or a sequence of experiments is by drawing a tree diagram. The idea is to draw, for each experiment, as many line segments as there are possible outcomes and to indicate each outcome by a brief symbol at the end of each line segment or branch. Then the outcomes of a second experiment are indicated by adding to each branch for the first experiment, as many branches as the number of possible outcomes for the second experiment, and so on. Each outcome of the whole sequence of the experiments will thus appear along a path, consisting of as many branches as there are experiments. This procedure is perhaps best illustrated with some examples.

Examples 4.2

1. A coin is tossed once and then a die is rolled once. Draw a tree diagram to indicate all possible outcomes. This is a sequence of two experiments.

1st. Tossing a coin has two possible outcomes. Writing H for heads and T for tails, we get a tree with two branches (see Figure 4.1).

Figure 4.1

2nd. Rolling a die has six possible outcomes. Indicating the face that turns up by the number of dots on the face, we get a tree with six branches for this part of the experiment (see Figure 4.2).

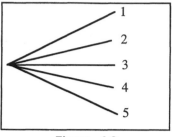

Figure 4.2

The second set of outcomes is possible with each outcome of the first set. Therefore, attaching the second tree to each branch of the first we get the complete tree diagram as shown in Figure 4.3.

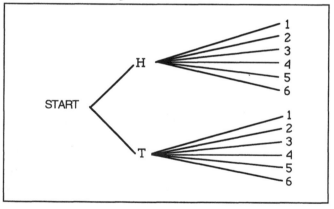

Figure 4.3

So, U = {H-1, H-2, H-3, H-4, H-5, H-6, T-1, T-2, T-3, T-4, T-5, T-6}

2.   A basket has ten oranges, one apple, and one lemon. Three fruits are taken out from the basket without replacement. Draw a tree diagram indicating all possible outcomes.

This is a sequence of three drawings: the first fruit, the second fruit, and the third fruit. Writing O for orange, A for apple, and L for lemon, we see that the first outcome can be either O, A, or L. If the first is O, then the second can be O, A, or L. If the first is A, then the second can only be O or L since there is only one apple. Similarly, if the first is L, then the second may be only O or A. In the same way, we get various cases for the third fruit [Figure 4.4.]

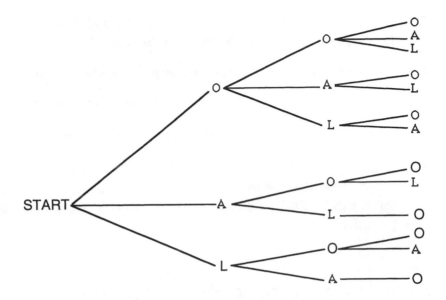

U = {OOO,OOA,OOL,OAO,OAL,OLO,OLA,AOO,AOL,ALO,LOO,LOA,LAO}

Figure 4.4

Exercises 4.1

For each of the following problems, draw a tree diagram indicating all the possible outcomes and then give the set of all possible outcomes.

1. A coin is tossed three times.

2. A basket has ten lemons and two oranges. Three fruits are taken out of the basket one after the other without replacement.

3. Change for 30 cents is to be given using quarters, nickels, and dimes.

4. There are four roads connecting city A to city B. A person wishes to make a trip from city A to city B and back such that he would not come back by the same road.

5. A person wishes to order a three course meal from a menu. The menu lists three soups: vegetable, tomato, and clam; two main courses: steak and chicken; and two desserts: ice cream and apple pie.

6. A committee of two consisting of a chairperson and a secretary is to be selected from three people.

7.   A basket has two oranges, two apples, and one lemon. Three fruits are taken from the basket one after the other without replacement.

8.   A coin is tossed until either a head comes up or two tails in a row occur.

9.   A coin is tossed until either a head comes up or three tails in a row occur.

10.   A coin is tossed until either a head comes up or four tails in a row occur.

11.   (a) A man has a dime, a nickel, and a penny in his pocket. He takes out two coins one after the other.

      (b) A man has a dime, a nickel, and a penny in his pocket. He takes out two coins.

2. THE FUNDAMENTAL PRINCIPLE OF COUNTING

If a coin is tossed once and a die is rolled once, then we have a sequence of two independent experiments. The first experiment has two possible outcomes and for each of those outcomes the second has six possible outcomes. Therefore, the total number of possible outcomes is 2×6. The total number of possible outcomes of a sequence of two experiments is the product of the number of possible outcomes of each of the two experiments -- the only requirement is that regardless of the first outcome, there are a constant number of possible outcomes for the second experiment. This principle forms the basis of most counting techniques.

> **If one experiment has exactly m possible outcomes, and for each of these outcomes, a second experiment has exactly n possible outcomes, then the sequence of the two experiments has exactly m × n possible outcomes.**

Examples 4.3

1.   A die is rolled twice. How many possible outcomes are there?

Solution. This is a sequence of two experiments, each of which has 6 outcomes. Therefore, the total number of outcomes for the sequence is $6 \times 6 = 36$.

2.   How many two digit numbers are there?

Solution. This can be set up as a sequence of two experiments - first, selecting the digit for the first place (9 possibilities), and second, selecting the digit for the second place (10 possibilities). Thus there are 9×10 or 90 two digit numbers.

3. There are 5 roads between New York and Boston. How many different ways can a person make a trip from Boston to New York and back if he does not wish to take the same road both ways?

Solution. He has 5 choices for the trip from Boston to New York. For each of these, he has 4 choices for the return trip. Hence the answer is 5 × 4 or 20.

The fundamental principle of counting can be applied to a sequence of more than two experiments. We illustrate this with the following examples:

4. A coin is tossed 3 times. How many possible outcomes are there ?

Solution. This is a sequence of three experiments. The first has 2 possible outcomes, for each of which the second has 2 possible outcomes, and for each of the first two, the third has two possible outcomes. So, the answer is: 2 × 2 × 2 = 8.

5. (a) How many three digit numbers can be written using the numerals 1, 2, 3, 4, 5, such that no digit occurs more than once?   60

    (b) How many of these numbers are less than 400 ?   14

Solution. (a) 5 × 4 × 3. To answer the second part, we note that for the number in the hundreds' place we have only three possibilities, 1, 2, and 3, so the answer to (b) is 3 × 4 × 3.

Multiplication versus Addition

The key phrase in the fundamental principle of counting is *for each*. If the second outcome does not occur 'for each' outcome of the first, the principle does not apply. Thus, if a coin is tossed once or a die is rolled once, we do not get a sequence of two experiments, but rather one or the other of the two given experiments. Therefore, the number of outcomes is 2 + 6, or 8, not 2 × 6, or 12. In general, if two experiments are connected by the phrase '*and*', then the number of possible outcomes is the product of the number of outcomes for each of the two experiments. If the experiments are connected by '*or*', then the number of possible outcomes is the sum of the number of outcomes possible for each experiment.

Exercises 4.2

1. A coin is tossed 5 times. How many possible outcomes are there?

2. There are 20 females and 10 males in a math class. How many handshakes will there be if each female shakes hand with each male?

3.   How many 5 digit numbers can be written using each of 1, 2, 3, 4, 5 only once? How many of these numbers will be less than 30,000 ?

4.   A state uses 2 letters followed by 3 numerals on the license plates. How many cars can be registered using this system? How should the system be changed if the number of cars increases 24 times?

5.   The telephone numbers within one area code consist of 7 numerals. How many different telephones can you have in one area? What is done if the number of telephones exceeds this number?

6.   The area code of a telephone number consists of a three digit number. How many different area codes can you have? What will happen if the number of area codes exceed this number?

7.   (a) A card is drawn from a deck of cards. How many different outcomes are possible?
      (b) Two cards are drawn one after the other and at random from a deck of playing cards. How many different outcomes are possible?

8.   (a) A die is rolled once or a card is drawn from a deck of cards. How many different outcomes are possible?
      (b) A die is rolled once and a card is drawn from a deck of cards. How many different outcomes are possible?

9.   (a) A coin is tossed twice and a die is rolled once. How many different outcomes are possible ?
      (b) A coin is tossed twice or a die is rolled once. How many different outcomes are possible ?

10.   (a) A coin is tossed once and two dice are rolled. How many different outcomes are possible?
      (b) A coin is tossed once or two dice are rolled. How many different outcomes are possible?

11.   A card is drawn from a deck of cards. If it is a heart, a coin is tossed once. Otherwise another card is drawn from the deck. How many different outcomes are possible?

12.   How many different ways can you answer a 10 question true/false test? How many ways can it be answered if no two consecutive answers are the same?

3. PERMUTATIONS

DEFINITION. An arrangement of n objects in a certain order is a *permutation* of the n objects.

We are first interested in finding the number of possible permutations of n objects. For example, consider the case $n = 2$ and let the two objects be denoted by a and b. Then there are two possible permutations of a and b, namely

a b   and   b a

For $n = 3$, let the objects be denoted by a, b, and c. Then there are six possible permutations given by (see Figure 4.5)

a b c   b a c   c a b

a c b   b c a   c b a

We can think of it as a sequence of three experiments: (1) selecting the object to go first, (2) selecting the object to go second, and (3) selecting the object to go last. We have three possibilities for the first object, a, b, or c. With each of these possibilities we have two possibilities for the second object. Thus, there are 3×2 possibilities for the first two objects in the sequence. Since there are only three objects, the third one is decided once the first two are chosen. In other words, there are $3 \times 2 \times 1$ possible permutations of 3 objects.

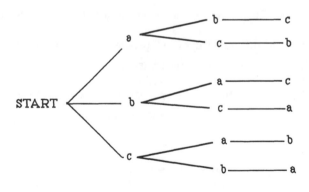

Figure 4.5

We can extend this argument for any number of objects. For $n = 4$, there will be 4 possibilities for the first, and for each of those there will be three possibilities for the second, and for each of the first two there will be two possibilities for the third, and with each of the first three, there will be only one choice remaining for the fourth. Thus, we have in all $4 \times 3 \times 2 \times 1$ possible permutations of 4 objects.

In general, for any n there will be

$$n \times (n-1) \times (n-2) \ldots \times 3 \times 2 \times 1$$

possible permutations for n objects. This number occurs very frequently. It also gets larger and larger as the value of n increases. This number is denoted by

n! (n factorial)

$$n! \;=\; n \times (n-1) \times (n-2) \ldots \times 3 \times 2 \times 1 \qquad\qquad (1)$$

For example, $4! = 4 \times 3 \times 2 \times 1 = 24$

$5! = 5 \times 4 \times 3 \times 2 \times 1 = 120.$

We have thus established the following theorem.

THEOREM 4.1.   *The number of possible permutations of n objects is n!.*

Examples 4.4

1.   How many different ways can 9 people be lined up for a group photograph?

Solution. This is the possible permutations of 9 objects. So, the answer is 9!.

2.   How many different ways can 9 people be lined up for a group photograph if one of them must always be at the center?

Solution. This time since one remains seated in the center it is the possible permutations of 8 objects. So, the answer is 8!.

We define

$$0! = 1 \qquad\qquad (2)$$

Let us now answer the question: how many permutations are possible for any three letters selected from the five letters a, b, c, d, e? This time there are 5 choices for the first letter and for each of these, 4 for the second, and for any of the first two, 3 for the third. Thus the answer is $5 \times 4 \times 3$ - this number is denoted by $_5P_3$ which means the number of permutations of 5 objects taken 3 at a time. In general, the number of permutations of n things taken r at a time is denoted by nPr or P(n, r).

We get the following formulas:

$$nPr = \underbrace{n(n-1)\ldots(n-r+1)}_{r \text{ factors}} \qquad (3)$$

$$nPn = n! \qquad (4)$$
$$nP_0 = 1 \qquad (5)$$

Examples 4.5

1. $_{10}P_3 = 10 \times 9 \times 8 = 720$

2. $_6P_4 = 6 \times 5 \times 4 \times 3 = 360$

3. How many 4 digit numbers can be written using any of the symbols 1, 2, 3, 4, 5, 6, and 7 only once?
    Solution. $_7P_4 = 7 \times 6 \times 5 \times 4 = 840.$

Exercises 4.3

1. Compute each of the following numbers:
    (a) 6!   (b) 0! + 1! + 2!   (c) 10! × 0!   (d) $_7P_3$
    (e) $_7P_7$ (f) $_{10}P_4$ (g) $_{100}P_0$ (h) $\dfrac{_{10}P_3}{3!}$

2. How many different ways can 7 people be lined up for a group photograph?

3. How many different ways can 7 people be lined up for a group photograph if one of them must always be at the center?

4. How many different ways can the 5 tires of a car be rotated? How many ways can this be done if the spare is too worn out to be used anywhere else?

5. There are 11 positions on a soccer team. How many different ways can these 11 positions be assigned to 11 of 15 players if each player can play in any position?

6. In how many different ways can 5 of 11 players be assigned to the 5 positions on a basketball team?

7. How many different three digit numbers can be written by using the numerals 1, 2, 3, 4, 5, 6, 7 if the same numeral cannot be used more than once in any number?

8. How many different five digit numbers can be written by using the numerals 1, 2, 3, 4, 5, 6, 7, 8, 9 if the same numeral cannot be used more than once in each number?

9. Two cards are drawn in succession from a deck of cards. How many different possibilities are there for the suits of the two cards?

10. At a freshman party, there were 17 males and 14 females. The band decided to play as many numbers as would be necessary for each male to dance with each female exactly once. How many numbers did the band play?

11. Four couples go to a movie. How many ways can they be seated on eight seats in a row if all the men must sit together and all the women must sit together?

12. Four students take their dates to a movie. How many ways can they be seated on eight seats in a row if each student must be seated next to his date?

4. COMBINATIONS

The number nCr

Suppose we are to select two letters from a, b, and c when it is immaterial which one is selected first. Then the possible selections are:

a, b   b, c   and   c, a

In this case the selections a, b and b, a are one and the same. This is an example of a combination of two things from three possible things. We say there are three different combinations of 2 selected from 3 or that the number of ways 2 things can be selected from 3 is 3.

DEFINITION. A selection of r objects taken from a set of n objects in any order is a combination. The number of possible combinations of r objects taken from n possible objects is denoted by

$$nCr \text{ (n choose r)} \quad \text{or} \quad C(n,r) \quad \text{or} \quad \binom{n}{r}$$

We shall use nCr since it is easier to write and many calculators use this notation on the key that performs this operation. A combination, unlike a permutation, does not involve order. For example:

$$_3P_2 = 6 \quad \text{but} \quad _3C_2 = 3.$$

This can be seen by considering three objects a, b, and c. Then

Permutations of 2 of a, b, and c are: a, b   b, a   b, c   c, b   c, a   a, c

Combinations of 2 of a, b, and c are: a, b   b, c   c, a.

THEOREM 4.2. $\quad nCr = \dfrac{nPr}{r!}$

Proof. The permutations of r from n is nPr. Since the number of permutations of r objects is r!, each combination of r objects appears r! times in nPr. Hence,

$$nCr \times r! = nPr.$$

Dividing both sides by r!, we get the result.

Corollary-1. $\quad nCr = \dfrac{n!}{(n-r)! \times r!}$

This is obtained by putting the value of nPr from formula (4) into the theorem.

The most useful formula for combinations is

$$nCr = \frac{n(n-1)\ldots(n-r+1)}{r!} \qquad (6)$$

which can be obtained by breaking up n! in corollary-1 as

$$n! = n(n-1)\ldots(n-r+1) \times (n-r)!.$$

Figure 4.8 indicates an easy way to remember formula (6).

$$nCr = \frac{\overbrace{n(n-1)\ldots(n-r+1)}^{r \text{ factors}}}{r!}$$

Figure 4.8

Examples 4.6

1. $\quad _{10}C_3 = \dfrac{(10 \times 9 \times 8)}{3!} = 120$

2. $\quad _{12}C_4 = \dfrac{(12 \times 11 \times 10 \times 9)}{4!} = 495$

Replacing r by n - r in corollary-1 of Theorem 4.2, we get the important formula

$$nCn\text{-}r \;=\; nCr \qquad\qquad (7)$$

Formula (7) is very important in the sense that it helps to simplify calculations. For example, $_{100}C_{98} = _{100}C_2 = \dfrac{100 \times 99}{2} = 4950.$

It also follows from the definition that

$$nCn \;\;=\;\; 1 \qquad\qquad (8)$$
$$nC_0 \;\;=\;\; nCn \;\;=\;\; 1 \qquad\qquad (9)$$

Examples 4.7

1. How many ways are there of forming a committee of 3 persons chosen from a group of 12 people?

Solution. Since a committee is to be formed by taking any 3 persons selected in any order from 12 possible people, it is like a combination of 3 things out of 12. So, the answer is $_{12}C_3 = (12 \times 11 \times 10) \div 3! = 220.$

2. A basketball league has 10 teams. How many games must be played so that each team gets to play every team exactly once?

Solution. For each game, we need 2 teams to be selected from 10 teams. Hence the answer is $_{10}C_2 = (10 \times 9) \div (2!) = 45.$

3. A coin is tossed 6 times.
    (a) How many outcomes are possible?
    (b) How many of the outcomes have 4 heads and 2 tails?
    (c) How many of the outcomes have at least 5 heads?
    (d) How many of the outcomes have at least 4 heads?
    (e) How many of the outcomes have at least 1 tails?

Solutions.   (a) $2 \times 2 \times 2 \times 2 \times 2 \times 2 = 2^6$

(b) $_6C_4 = _6C_2$

(c) $_6C_5 + _6C_6 = _6C_1 + _6C_0$

(d) $_6C_4 + {}_6C_5 + {}_6C_6 = {}_6C_2 + {}_6C_1 + {}_6C_0$

(e) This is equivalent to saying that the outcome is not all tails because all tails can occur in 1 way. The answer is $2^6 - 1$.

Permutation Vs. Combination

In many problems, it is often not easy to decide whether to use permutation or combination. One simple criterion is to ask the question "Does changing the order of the objects result in a different outcome?" If the answer is "yes", then it is a permutation. If the answer is "no", then it is a combination. For example, consider:

There are 8 different fruit juices. A punch is to be made by mixing equal amounts of any five of the fruit juices. The flavor of the punch depends on the order in which the fruit juices are mixed. How many different kinds of punch may be made from the 8 fruit juices?

Here, the answer to the question: 'Does changing the order of mixing the juices result in a different punch?' is "yes". So, it is a case of permutation and the answer is $_8P_5$.

However, if the problem is changed to:

There are 8 different fruit juices. A punch is to be made mixing equal amounts of any five of the fruit juices. If the flavor of the punch does not depend on the order in which the fruit juices are mixed, how many different kinds of punch may be made from the 8 fruit juices?

In this case, the answer to the question: 'Does changing the order of mixing the juices result in a different punch?' is "no". Therefore, it is a case of combination and the answer is $_8C_5$.

Exercises 4.4

1. Compute each of the following numbers:

   (a) $_5C_2$ (b) $_7C_2$ (c) $_{12}C_4$ (d) $_{10}C_0$

   (e) $_{100}C_{100}$ (f) $_{50}C_{49}$ (g) $_{200}C_{198}$ (h) $_3C_0 + {}_3C_1 + {}_3C_2 + {}_3C_3$

2. How many ways can a committee of 7 be chosen from a group of 10 people ?

3. A club has 25 members. How many ways are there of forming an entertainment committee of 3 members and an executive committee of 4 members, if

   (a) Any member can serve on both committees?

   (b) No one can serve on more than one committee?

4.   How many different ways can an 11 player soccer team be selected from a group
      of 15 possible players?

5.   A basketball league has 15 teams. How many games will the league have if
      (a) Each team plays every team exactly once?   225
      (b) Each team plays every team exactly twice?

6.   20 generals of the Chinese Red Army meet to celebrate their victory in a battle.
      As part of the celebration, each general hugs every other general and a
      photographer takes a picture of each hug. How many pictures did the
      photographer take?

7.   A coin is tossed 5 times.
      (a) How many outcomes are possible?
      (b) How many of the outcomes have 3 heads and 2 tails?
      (c) How many of the outcomes have 2 heads and 3 tails?
      (d) How many of the outcomes have at least 3 heads?
      (e) How many of the outcomes have at least 1 tails?

8.   A student takes a 10 question T-F test. How many different ways can he
      answer the test so that
      (a) The answer has 6 T's and 4 F's ?
      (b) The answer has 4 T's and 6 F's?
      (c) The answer has 5 T's and 5 F's?
      (d) How many of the outcomes have at least 8 T's?

9.   There are 10 males and 12 females in a coed dorm.
      (a) How many ways can a committee of 2 males and 3 females be formed?
      (b) How many ways can an entertainment committee of 3 males and 3 females,
          and an executive committee of 2 males and 2 females be formed if no one
          can serve on both committees?

10.   Three committees are to be formed from a group of 11 students. The first
      committee has 2 members, the second has 3 and the third committee has 4
      members. How many ways can the committees be formed if
      (a) Anyone can serve on more than one committee?
      (b) No one can serve on more than one committee?

11.   Ten persons form a group to play tennis in the winter. They book 2 courts once
      a week so that any 8 of them can play on a particular week. If they wish that

every possible combination of 8 players gets to play exactly once, how many weeks should they book the courts for?

12. There are 10 different fruit juices. A punch is to be made mixing equal amounts of any five of the fruit juices. The flavor of the punch depends on the order in which the fruit juices are mixed. How many different kinds of punch may be made from the 10 fruit juices?

13. In problem 12, how many different kinds of punch can be made if the flavor of the punch does not depend on the order in which the fruit juices are mixed?

14. An artist makes different colors by combining in any order equal amounts of three of the colors in the rainbow. How many different colors can he or she create if the order in which the three colors are mixed is immaterial?

15. If a coin is tossed 10 times, how many possibilities result in at least 8 heads?

16. A quarter, a dime, a nickel and a penny are tossed. How many ways can any two of them come up heads?

17. An ice-cream parlor offers 10 different flavors of ice-cream. How many different two scoop cones can a customer order?

CHAPTER TEST 4

1. Compute each of the following:

   (a) $_7P_5$ (b) $_{10}C_9$ (c) $_{100}C_{98}$

   (d) $0! + 1! + 4!$ (e) $_9C_0 + {_9}C_1 + {_9}C_9$

2. A basket has 7 apples and 2 oranges. Three fruits are taken out of the basket one after the other without replacement. Draw a tree diagram indicating all possible outcomes and then give the set of all possible outcomes.

3. The license plates of a state are made using three letters followed by three numerals. How many different cars can be registered using this system?

4. (a) How many different outcomes are possible if three dice are rolled once?

   (b) Give the number of possible outcomes if either a die is rolled once or a coin is tossed once.

5.   A coin is tossed 5 times. How many different ways can it come up heads at least 3 times?

6.   There are 18 females and 12 males in a math class.

   (a) How many handshakes will there be if each female shakes hand with each male?

   (b) How many handshakes will there be if every student shakes hand with every student in the class?

7.   A soccer team has 2 goalies, 7 defense men, and 8 forwards. How many ways can 1 goalie, 5 defense men, and 5 forwards be selected from this team?

8.   How many different six digit numbers can be formed using the numerals 1, 2, 3, 4, 5, 6, 7, 8, 9, if the same numeral is not used more than once in any number? How many of these numbers will be less than 600,000?

9.   (a) A vegetable soup is made by boiling together six different kinds of vegetables. How many different kinds of soups can be prepared if twelve different kinds of vegetables are available?

   (b) How many different 4 layer cakes can be made by using any four of six different ingredients that may be used for a layer?

10.   A student takes a 10 question T-F test. How many different ways can he answer the test so that

   (a) The answer has 7 T's and 3 F's ?

   (b) The answer has 8 T's and 2 F's?

   (c) The answer has 2 T's and 8 F's?

   (d) How many of the outcomes have at least 9 T's?

Chapter 5

PROBABILITY THEORY

The Very Guide of Modern Life

Probability is the mathematical way of analyzing situations that involve chance. It is a logical way of predicting the chances of something happening. One may ask "Why study probability theory?" We list a few things in answer to that question.

(1) In order to make decisions ahead of time, we need to know what can happen or what is most likely to happen.

(2) We need to estimate the size of a population as in estimating the number of people likely to come to a reception or in deciding if a rare animal, bird, or fish faces extinction.

(3) To figure out the odds before we act as in gambling, cards, and other games.

(4) To measure our expectations.

(5) To study random phenomena.

(6) Last but not the least, we give a news item to emphsize that if you know a little bit of probability theory then you will perhaps not gamble like many people do.

> "A Stamford woman who lost heavily at the nearby Foxwoods Resort Casino over the weekend killed herself shortly after leaving the gaming tables, authorities said yesterday. The body of Laura Grauer, 38, was found floating in the Thames River Sunday morning by a man walking his dog near Fort Shantok State Park. Her death was ruled a suicide by the chief state medical examiner. State police and relatives confirmed that Grauer left the casino early Sunday morning after gambling the maximum amount on her credit card. (AP)"[The Providemce Journal-Bulletin, March 8, 1996].

It may seem absurd that the mathematical way of reasoning can be applied to something so unpredictable as "chance", yet probability theory has endless applications in arts, sciences, gambling, government, and business. These days, probability is regarded as the very guide of modern life. Probability theory perhaps provides the best guidance in a situation where one has to make a decision in the face of uncertainties.

The main purpose of probability theory is to assign a number (probability measure) to an outcome in such a manner that the assigned number reflects the likelihood of that outcome happening. For example, if a coin is tossed, we say that the probability of getting a head is 1/2. This may be interpreted to mean that if a coin is tossed a certain number of times, then we can predict that half of the time heads will turn up. This prediction may not come true, but the method behind this prediction is very logical. We shall at first consider how to assign probability to an event.

1. PROBABILITY MEASURE

Events. Let us recall that the sample space U of an experiment is the set of all possible outcomes of the experiment. Now we use U and set language to clarify the meaning of an event.

DEFINITION. An *event* is a subset of the sample space U of an experiment. An event is thus a set of outcomes contained in the sample space.

Examples 5.1

1. A coin is tossed twice. Find the event E: 'it comes up heads at least once'. In this case, E = {HT, HH, TH}.

2. A die is rolled once. Find the event E: 'an odd number turns up'. If we denote this event by E, then E = {1, 3, 5}

3. A card is drawn from a deck of playing cards. Find each of the following events:
   E: 'it is a face card'.
   F: 'the card is neither black nor red'.

   Solution.   E = {♣K, ♣Q, ♣J, ◇K, ◇Q, ◇J, ♡K, ♡Q, ♡J, ♠K, ♠Q, ♠J}
                  F = ∅.

4. A die is rolled twice. Find each of the following events:
   E: 'the sum of the two outcomes is 11'
   F: 'it comes up with a 6 at least once'

Solution. In this case, U will have 36 elements. The elements of U are listed in Figure 5.1 below. Note that (1, 2) refers to the outcome of 1 on the first roll and 2 on the second roll, and so on. So,

E = {(6, 5), (5, 6)} and

F = {(6, 1), (6, 2), (6, 3), (6, 4), (6, 5), (6, 6), (1, 6), (2, 6), (3, 6), (4, 6), (5, 6)}.

Note: (6, 6) appears in the list only once.

(1, 1)	(1, 2)	(1, 3)	(1, 4)	(1, 5)	(1, 6)
(2, 1)	(2, 2)	(2, 3)	(2, 4)	(2, 5)	(2, 6)
(3, 1)	(3, 2)	(3, 3)	(3, 4)	(3, 5)	(3, 6)
(4, 1)	(4, 2)	(4, 3)	(4, 4)	(4, 5)	(4, 6)
(5, 1)	(5, 2)	(5, 3)	(5, 4)	(5, 5)	(5, 6)
(6, 1)	(6, 2)	(6, 3)	(6, 4)	(6, 5)	(6, 6)

Figure 5.1

Exercises 5.1

In doing the following exercises, it may be easier at times to draw tree diagrams.

1. A coin is tossed twice. Give
   (a) the sample space,
   (b) the event: 'it doesn't come up heads even once'.

2. A coin is tossed and then a die is rolled. Give
   (a) the sample space,
   (b) the event: 'head and a multiple of 3 is rolled'.

3. A coin is tossed three times. Give
   (a) the sample space,
   (b) the event: 'exactly twice it comes up heads',
   (c) the event: 'it comes up heads at least once'.

4.   A coin is tossed four times. Give

(a) the sample space,

(b) the event: 'exactly twice it comes up heads',

(c) the event: 'it comes up heads at least once'.

5.   A die is rolled twice. Find each of the following events:

(a) 'the sum of the two numbers that turn up is 10',

(b) 'the sum of the two numbers that turn up is 7',

(c) 'the sum of the two numbers that turn up is at least 10',

(d) 'it comes up with a 4 at least once'.

6.   A die is rolled twice. Find each of the following events:

(a) 'the sum of the two numbers that turn up is 5',

(b) 'the sum of the two numbers that turn up is 8',

(c) 'the sum of the two numbers that turn up is at least 11',

(d) 'the sum of the two numbers that turn up is an odd number'.

7.   An urn contains three black balls and a white one. Two balls are taken out one after the other without replacing them. Find each of the following:

(a) the sample space,

(b) the event: 'at least one ball is black',

(c) the event: 'at least one ball is white',

(d) the event: 'both balls are white'.

8.   A card is drawn from a deck of playing cards. Find each of the following events: E: 'it is an honor card',   F: 'it is a black face card'.

Weights.

Weights. The first step in determining the probability of an event is to find the corresponding sample space U. Then each element of U is assigned a number, called its *weight*, such that the sum of the weights is 1.

The *probability measure* of an event E, denoted by Pr[E], is the sum of the weights of the elements in E. It follows immediately that

1.   If $E = U$ then $Pr[E] = 1$.

2.   If $E = \varnothing$ then $Pr[E] = 0$.

3.   If E is a proper subset of U, then Pr[E] is a proper fraction.

Note that a proper fraction is less than one and is positive.

The assignment of a weight to each possible outcome of an experiment usually depends on known, given, or factual information. Once these weights have been assigned, it is very easy to determine the probability of an event.

Examples 5.2

1. A letter is chosen from the word 'mathematics'. Find the probability that it is a 't'.

Solution. In this case, U = {m, a, t, h, e, i, c, s}. To assign weight to each of these possibilities, we note that each of the letters m, a, and t occurs twice; so the weights of these 3 letters must be twice that of the others. The numerators of the weights are as shown in Figure 5.2.

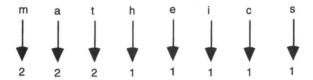

Figure 5.2

Since each weight must be a positive fraction less than 1, and since the sum of the weights must be 1, we find a common denominator for the weights. This will happen if the sum of the numerators equal the denominator. So, the denominator is equal to $2 + 2 + 2 + 1 + 1 + 1 + 1 + 1 = 11$.

Hence if E = {t} then Pr [E] $= \dfrac{2}{11}$

2. A die is loaded in such a way that the probability of each face is proportional to the number of dots on that face. What is the probability of getting an even number on one throw?

Solution. This means that the face 2 has twice as much probability as the face 1, the face 3 has three times the probability as the face 1, and so on. The numerator of the weights of each face can be taken as shown in Figure 5.3. Since

$1 + 2 + 3 + 4 + 5 + 6 = 21$, the denominator of the weights is 21.

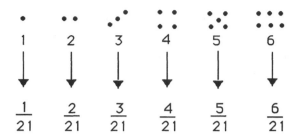

Figure 5.3

Since E = {2, 4, 6},

$$\text{Pr}[E] = \frac{2}{21} + \frac{4}{21} + \frac{6}{21} = \frac{12}{21} = \frac{4}{7}.$$

Exercises 5.2

1.   A letter is chosen from the word 'exercises'. What is the probability that it is an 'e'? What is the probability that the letter is a vowel?

2.   A letter is chosen from the word 'multicultural'. What is the probability that it is an 'u'? What is the probability that the letter is a vowel?

3.   An urn contains three black, two white, and two red balls. A ball is drawn from this urn. What is the probability that
      (a)   the ball is white?
      (b)   the ball is yellow?
      (c)   the ball is yellow or black?

4.   A jar has 3 cherry, 1 lemon and 6 orange gumdrops. A child picks at random a gumdrop from the jar. What is the probability that
      (a)   he has picked a cherry gumdrop?
      (b)   he has picked a lemon gumdrop?
      (c)   he has picked a grape gumdrop?
      (d)   he has picked a cherry or a grape gumdrop?

5.   A die is loaded in such a way that the probability of each face is proportional to the number of dots on that face (see 2 of Examples 5.2). The die is rolled once. What is the probability that the face that turns up has more than 3 dots?

6.   A defective coin is such that each time it is tossed the probability of it coming up 'heads' is twice that of it coming up 'tails'. The coin is tossed once. What is the probability that it comes up 'tails' ?

7.   An expectant mother knows that she will give birth to twins. What is the probability that
(a) both of them are of the same sex?
(b) they are of the opposite sex?
(c) they are both girls?

8.   A child had 1 cherry, 5 grape and 8 orange gumdrops in his pocket. He carelessly lost one gumdrop. What is the probability that
(a) the cherry gumdrop was lost?
(b) an orange gumdrop was lost?

9.   An expectant mother knows that she will give birth to triplets. What is the probability that
(a) she would give birth to two girls and one boy?
(b) she would give birth to three girls?

10.   A die is loaded in such a way that the face with an odd number of dots has twice as much probability than that with an even number of dots. The die is rolled once. What is the probability that an odd number turns up?

11.   Two dice are rolled once. List the possible values for the sum of the two numbers that turn up and give the probability of each sum.

12.   Two dice are rolled once. Find the probability of each of the following:
E: 'the sum of the two numbers that turn up is 7',
F: 'the sum of the two numbers that turn up is 8'.

13.   Show that the four children in a family are more likely to consist of three of one sex and one of another than to consist of two boys and two girls.

14.   List the possible sex distributions for a family of five children.

2. EQUIPROBABLE MEASURE

If the sample space of an experiment is a finite set U and if each outcome of U is equally likely, then the weight assigned to each outcome is $\frac{1}{n(U)}$. In this case, for any event E,

$$Pr[E] = \frac{n(E)}{n(U)} \qquad\qquad (1)$$

where n(E) denotes the number of elements in E and n(U) denotes the number of elements in U. The probability measure determined in this manner is an *equiprobable measure*.

From the formula n(E)/n(U), we can conclude that in solving problems using equiprobable measure, it is not really necessary to find U or the subset E, but it is sufficient to find the number of elements in each of the sets U and E. Therefore, the counting techniques and especially the method of counting permutations and combinations are very useful in obtaining equiprobable measure.

The term *selected at random* indicates that each element of the sample space is to be given the same weight. That is, each possible outcome has the same probability. Thus, 'a card is selected (or drawn) at random from a deck of playing cards' means that it can be any one of the 52 cards and each one has probability 1/52. Therefore, equiprobable measure must be used whenever it is given that the selection is at random.

Examples 5.3

1.   A card is drawn at random from a deck of playing cards. Let

   E: 'the card drawn is a heart'.

   F: 'the card drawn is an honor card'.

   Find each of the following:

   (a) n(U)   (b) n(E)   (c) Pr[E]   (d) n(F)

   (e) Pr[F]   (f) E ∩ F   (g) n(E ∩ F)   (h) Pr[E∩ F]

Solution.   (a) 52   (b) 13   (c) $\frac{13}{52} = \frac{1}{4}$ (d) 16

   (e) $\frac{16}{52} = \frac{4}{13}$ (f) {♡A, ♡K, ♡Q, ♡J}   (g) 4

   (h) $\frac{4}{52} = \frac{1}{13}$

2. Two dice are rolled once. Find the probability that
    (a) 'the sum of the two numbers that come up is 6'.
    (b) 'at least one die comes up with the number 6'.

Solution.   This experiment is very similar to the experiment of rolling a die twice. In this case also, we can use the outcomes listed in Figure 5.1. The only difference is that the outcome (1, 5) will now mean 1 on the first die and 5 on the second die, and so on. So, $n(U) = 36$.

(a) If E : 'the sum of the two numbers that come up is 6', then
$$E = \{(1, 5), (5, 1), (4, 2), (2, 4), (3, 3)\},$$

$n(E) = 5$, and $\Pr[E] = \dfrac{n(E)}{n(U)} = \dfrac{5}{36}$.

(b) If F : 'at least one die comes up with a 6', then

$$F = \{(6,1),(6,2),(6,3),(6,4),(6,5),(6,6),(1,6),(2,6),(3,6),(4,6),(5,6)\}.$$

So, $n(F) = 11$ and $\Pr[F] = \dfrac{n(F)}{n(U)} = \dfrac{11}{36}$.

3. Two cards are drawn one after the other and at random from a deck of cards. What is the probability that E: 'they are both spades'?

Solution. In this case, it is necessary to think of the first card and the second card. So,
$$n(U) = {}_{52}P_2 = 52 \times 51$$

$$n(E) = 13 \times 12$$

Therefore, $\Pr[E] = \dfrac{13 \times 12}{52 \times 51} = \dfrac{1}{17}$.

4. Three cards are drawn at random from a deck of cards. What is the probability that they are all hearts (E)?

Solution.   The experiment is selecting 3 from 52. So,

$$n(U) = {}_{52}C_3 = \frac{52 \times 51 \times 50}{3!}$$

$$n(E) = \frac{13 \times 12 \times 11}{3!} \quad \text{and} \quad \Pr[E] = \frac{13 \times 12 \times 11}{52 \times 51 \times 50} = \frac{11}{850}.$$

Exercises 5.3

1.   A card is drawn at random from a deck of playing cards. Find the probability of each of the following events:

        (a)   E: 'the card drawn is a 6'

        (b)   F: 'the card drawn is a face card.'

2.   A card is drawn at random from a deck of playing cards. Let

        E: 'the card drawn is an honor card'

        F: 'the card drawn is a red card.'

Then find each of the following:

(a) n(U)	(b) n(E)	(c) Pr[E]	(d) n(F)
(e) Pr[F]	(f) E ∩ F	(g) n(E ∩ F)	(h) Pr[E ∩ F]

3.   A card is drawn at random from a deck of playing cards. Let

        E: 'the card drawn is not a diamond',

        F: 'the card drawn is a face card.'

Then find each of the following:

(a) n(U)	(b) n(E)	(c) Pr[E]	(d) n(F)
(e) Pr[F]	(f) E ∩ F	(g) n(E ∩ F)	(h) Pr[E ∩ F]

4.   A card is drawn at random from a deck of playing cards. Let

        E: 'the card drawn is not a face card',

        F: 'the card drawn is an honor card.'

Then find each of the following:

(a) n(U)	(b) n(E)	(c) Pr[E]	(d) E ∩ F
(e) n(E ∩ F)	(f) Pr[E ∩ F]		

5.   Two dice are rolled once. Find the probability of each of the following events:

        (a)   'the sum of the two numbers that come up is 8'.

        (b)   'at least one of the numbers that come up is 5'.

6.   Two dice are rolled once. Find the probability that

        (a)   the sum of the two numbers that come up is even,

        (b)   at least one of the numbers that come up is 3.

7.   A die is rolled twice. Find the probability of each of the following events:

    (a)   'it comes up with 1 each time'.

    (b)   'it comes up with 1 or 2 each time'.

8.   A die is rolled twice. Find the probability of each of the following events:

    (a)   'it comes up with an odd number each time'.

    (b)   'it comes up with 1 or 6 each time'.

9.   Two cards are drawn one after the other and at random from a deck of playing cards. Let   E: 'they are both diamonds',   F: 'both cards are kings'. Then find each of the following:

    (a) n(U)   (b) n(E)   (c) Pr[E]   (d) n(F)

    (e) Pr[F]   (f) $E \cap F$ (g) $n(E \cap F)$ (h) $Pr[E \cap F]$

10.   Two cards are drawn one after the other and at random from a deck of playing cards.   Let E: 'they are both red', F: 'both cards are face cards'. Then find

    (a) n(U)   (b) n(E)   (c) Pr[E]   (d) n(F)

    (e) Pr[F]   (f) $E \cap F$ (g) $n(E \cap F)$ (h) $Pr[E \cap F]$

11.   A coin is tossed 3 times. Let

        E: 'it will turn up heads exactly twice'

        F: 'it will turn up heads all three times'

        G: 'it will turn up tails at least once'.

    Then find each of the following:

    (a) n(U)   (b) n(E)   (c) Pr[E]   (d) n(F)

    (e) Pr[F]   (f) n(G)   (g) Pr[G]

12.   A coin is tossed 4 times. Let

        E: 'heads turn up exactly three times'

        F: 'heads turn up all four times'

        G: 'tails turn up at least once'.

    Then find each of the following:

    (a) n(U)   (b) n(E)   (c) Pr[E]   (d) n(F)

    (e) Pr[F]   (f) n(G)   (g) Pr[G]

13.   A coin is tossed 8 times. Find the probability of each of the following events:

        E: 'heads turn up exactly 4 times'

        F: 'heads turn up at least 6 times'.

14.   A coin is tossed 10 times. Find the probability of each of the following events.

        E: 'heads turn up exactly 7 times'

        F: 'heads turn up at least 9 times'.

15.   Three cards are drawn at random from a deck of cards. Find the probability that they are all face cards.

16.   Three cards are drawn at random from a deck of cards. Find the probability that they are all honor cards.

17.   Two red balls and two white balls are to be placed two each at random in two boxes. Give

    (a) the sample space,

    (b) the probability that both red balls will be in the same box.

18.   There are three persons in a room. Find the probability that no two of them were born in the same month.

19.   Two dice are rolled. Find the probability of each of the following events:

    (a) 'both dice would come up with the same number'.

    (b) 'the numbers that come up are two consecutive numbers'.

20.   A die is rolled three times. Find the probability of each of the following events.

      (a)   'it comes up with 6 each time'

      (b)   'it comes up with an even number each time'.

21.   Three dice are rolled once. List the possible values for the sum of the three numbers that turn up and give the probability of each sum. What value or values for the sum have the highest probability? What value or values for the sum have the lowest probability?

*22.   (Duke of Tuscany's problem) Three dice are rolled. Show that a sum of 10 is more likely to happen than a sum of 9.

3. PROPERTIES OF PROBABILITY MEASURE

We shall soon discuss various methods of computing probabilities of different kinds of events, but before we do so, it is necessary to understand a few basic properties of the probability measure. Since the probability of an event is defined in terms of sets, it is possible to apply results and ideas of set theory to probability.

THEOREM 5.1. *Let E and F be two events in the same sample space. Then*

$$Pr[E \cup F] = Pr[E] + Pr[F] - Pr[E \cap F] \qquad (2)$$

Proof: It follows from the definitions of probability measure, union of sets, and intersection of sets that

Pr[E ∪ F] = Sum of the weights of the elements in E

+ sum of the weights of the elements in F

- sum of the weights of the elements in E ∩ F

= Pr[E] + Pr[F] - Pr[E ∩ F].

DEFINITION. Two events E and F are *mutually exclusive* if E ∩ F = ∅, that is, if

$$Pr[E \cap F] = 0.$$

From theorem 5.1, it follows that

THEOREM 5.2. *If E and F are mutually exclusive events, then*

$$Pr[E \cup F] = Pr[E] + Pr[F].$$

The negation of an event E may be thought of as $\overline{E}$ (complement of event E), and we can apply the knowledge of the complement of a set to detrmine Pr[$\overline{E}$] . From the definition of probability measure, it follows that

$$Pr[E] + Pr[\overline{E}] = Pr[E \cup \overline{E}] = Pr[U] = 1.$$

Since E ∪ $\overline{E}$ = U and E ∩ $\overline{E}$ = ∅, we get

THEOREM 5.3.   *For any event E,*

$$Pr[\,\overline{E}\,] = 1 - Pr[E] \qquad\qquad (3)$$

Examples 5.4

1.   A card is drawn at random from a deck. Find the probability that the card drawn is not a face card.

Solution.   Here if we regard $\underline{E:}$ the card drawn is a face card, then
$\overline{E}$: the card drawn is not a face card.

We need to determine $Pr[\,\overline{E}\,]$. We use formula (3).

$$Pr[\,\overline{E}\,] = 1 - Pr[E] = 1 - \frac{3}{13} = \frac{13}{13} - \frac{3}{13} = \frac{10}{13}.$$

2.   A die is rolled once. Let E: 'the die comes up with 6',

F: 'the die comes up with an even number'.

Then find each of the following:

(a) $Pr[\,\overline{E}\,]$ (b) $Pr[E \cap F]$,   (c) $Pr[E \cup F]$

(d) $Pr[\,\overline{E}\, \cap F]$ (e) $Pr[\,\overline{E}\, \cup F]$

Solution.   (a) Since $Pr[E] = \frac{1}{6}$, $Pr[\,\overline{E}\,] = 1 - \frac{1}{6} = \frac{5}{6}$

(b) $Pr[F] = \frac{3}{6}$, $Pr[E \cap F] = \frac{1}{6}$

(c) $Pr[E \cup F] = \frac{1}{6} + \frac{3}{6} - \frac{1}{6} = \frac{3}{6}$

(d) $\overline{E} \cap F = \{2, 4\}$.   Therefore, $Pr[\,\overline{E} \cap F] = \frac{2}{6} = \frac{1}{3}$.

(e) $Pr[\,\overline{E} \cup F]$ $= Pr[\,\overline{E}\,] + Pr[F] - Pr[\,\overline{E} \cap F]$

$$= \frac{5}{6} + \frac{3}{6} - \frac{2}{6} = 1.$$

Exercises 5.4

1. A card is drawn at random form a deck of cards. Let

    E: 'the card drawn is a heart'

    F: 'the card drawn is a king'.

    Then find each of the following:

    (a) Pr[E]   (b) Pr[$\overline{E}$]   (c) Pr[F]

    (d) Pr[$\overline{F}$]   (e) Pr[E ∩ F]   (f) Pr[E ∪ F]

    (g) Pr[$\overline{E}$ ∩ F]   (h) Pr[$\overline{E}$ ∪ F]

2. A card is drawn at random from a deck of cards. What is the probability that it is a diamond or a face card?

3. A card is drawn at random from a deck of cards. Let

    E: 'the card drawn is an honor card' (ace, king, queen, or jack)

    F: 'the card drawn is red'.

    Find each of the following:

    (a) Pr[E]   (b) Pr[$\overline{E}$]   (c) Pr[E ∪ F]   (d) Pr[$\overline{E}$ ∪ $\overline{F}$]

4. A card is drawn at random from a deck of cards. Let

    E: 'the card drawn is a spade'

    F: 'the card drawn is a face card'.

    Find each of the following:

    (a) Pr[E]   (b) Pr[$\overline{E}$]   (c) Pr[E ∪ F]   (d) Pr[$\overline{E}$ ∪ $\overline{F}$]

5. A die is rolled once. Let

    E: 'the die comes up five'

    F: 'the die comes up with an odd number'.

    Find each of the following:

    (a) Pr[E ∪ F]   (b) Pr[$\overline{F}$]   (c) Pr[E ∩ $\overline{F}$]   (d) Pr[E ∪ $\overline{F}$]

6. A die is rolled once. Let

    E: 'the die comes up with an even number'

    F: 'the die comes up with a number greater than 3'.

Find each of the following:

(a) Pr[E] (b) Pr[E ∩ F] (c) Pr[$\overline{F}$]

(d) Pr[$\overline{E}$ ∪ $\overline{F}$] (e) Pr[E ∪ $\overline{F}$]

(f) Are E and F mutually exclusive?

7. Two dice are rolled once. Let

E: 'the sum of the two numbers that come up is 7'

F: 'at least one of the two numbers that come up is a 3'.

Find each of the following:

(a) F (b) Pr[F] (c) Pr[$\overline{F}$]

(d) Pr[$\overline{E}$] (e) Pr[E ∩ F] (f) Pr[E ∪ F]

(g) Pr[$\overline{E}$ ∩ $\overline{F}$] (h) Pr[$\overline{E}$ ∪ $\overline{F}$]

8. Two dice are rolled once. Let

E: 'the sum of the two numbers that come up is 8'

F: 'at least one of the the numbers that come up that come up is a 4.'

Find each of the following:

(a) F (b) Pr[F] (c) Pr[$\overline{F}$]

(d) Pr[$\overline{E}$] (e) Pr[E ∩ F] (f) Pr[E ∪ F]

(g) Pr[$\overline{E}$ ∩ $\overline{F}$] (h) Pr[$\overline{E}$ ∪ $\overline{F}$]

9. Let E and F be two events in the same sample space such that Pr[E] = 1/2, Pr[F] = 4/10, and Pr[E ∩ F] = 1/10. Find each of the following:

(a) Pr[$\overline{F}$] (b) Pr[E ∪ F] (c) Pr[$\overline{E}$ ∩ $\overline{F}$]

10. Let E and F be two events in the same sample space such that Pr[E] = 2/3, Pr[F] = 1/4, and Pr[E ∩ F] = 1/6. Find each of the following:

(a) Pr[$\overline{F}$] (b) Pr[E ∪ F] (c) Pr[$\overline{E}$ ∩ $\overline{F}$]

11. A coin is tossed 8 times. Find the probability of each of the following events:

E: 'it turns up heads at least once'

F: 'heads turn up at least twice'.

12.   A coin is tossed 10 times. Find the probability of each of the following events:

E: 'it turns up heads at least once'

F: 'heads turn up at least twice'.

13.   Two cards are drawn one after the other and at random from a deck of cards. Find the probability that neither card is a diamond.

14.   Two cards are drawn one after the other and at random from a deck of cards. Find the probability that they are both honor cards.

15.   Two cards are drawn one after the other and at random from a deck of cards. Let

E: 'both cards are hearts'

F: 'the first card is a face card'.

Find each of the following:

(a) Pr[E]   (b) Pr[F]   (c) Pr[E ∩ F]   (d) Pr[E ∪ F]

16.   Two cards are drawn one after the other and at random from a deck of cards. Let

E: 'both cards are diamond'

F: 'the first card is a king'. Find each of the following:

(a) Pr[F]   (b) Pr[E ∩ F]   (c) Pr[E ∪ F]

17.   Two cards are drawn one after the other and at random from a deck of cards. Let

E: 'both are honor cards'

F: 'the first card is a heart'. Find each of the following:

(a) Pr[F]   (b) Pr[E ∩ F]   (c) Pr[E ∪ F]

Random Phenomena

You must be wondering why we've been working with all these problems about tossing coins, rolling dice, and dealing cards, all of which are essentially gambling tools. Are we trying to train you to be gamblers? It is not so -- all of these are problems to illustrate the theory of probability and describe many natural phenomena known as Random Phenomena. We wish to explian what we mean by a random phenomenon.

When we say that if we toss a coin, $Pr[H] = 1/2$, $Pr[T] = 1/2$, we mean two things:

(I)   First, any time we toss a coin, Pr[H] = 1/2 and Pr[T] = 1/2, no matter what happened in the past. Thus, if we toss a coin once and it comes up heads, then we should not expect that the next time we toss the coin it would come up tails. Similarly, even if the coin comes up heads three times in a row, we cannot hope that the fourth time it has to come up tails - the probability of getting tails in the fourth time is still 1/2, no matter what happened in the previous three throws.

(II)   Second, in the long run the ratio of times the coin comes up heads to the number of tosses will be closer and closer to 1/2 as we increase the number of tosses.

(I) and (II) actually characterize what we mean by a random phenomenon. In general, a phenomenon is random if it has the following two characteristics:

(a)   The outcomes are uncertain and

(b)   the relative frequency of each possible outcome in the long run is exactly or nearly the same fraction.

Like tossing a coin, rolling a die gives a random phenomenon because any time we roll a die we do not know which of the six faces is going to turn up (characteristic (a)). We do know that if we roll the die a large number of times the relative frequency of each face turning up will be 1/6 or a number close to 1/6 (characteristic (b)).

It is the opinion of the experts that most natural phenomena are random. Take, for example, the sex of a baby to be born. It happens like this: The egg of a female contains a single sex chromosome   X. The male sperm may contain either a single   X chromosome or a single Y chromosome. If a sperm with an X cell fertilizes the female egg cell then the offspring is XX (or female), and if the egg is fertilized by a Y cell then the offspring is of type XY (or male). Therefore, the sex of the offspring depends on the proportion of cells containing X or Y chromosomes in the male sperm. Since about half the sperm cells contain X chromosomes and the other half contain Y chromosomes, the probability that a baby is female is 1/2 and the probability that it is male is also 1/2. If a couple has more girls than boys there is nothing surprising. But it can be safely predicted that about half the babies born in a specified period of time will be males and the other half females.

So, your sex has been determined by a process that can be characterized as random. According to Quantum theory, at the subatomic level the natural world is random. Atomic particles released in fission also move in a random manner. Random phenomena abound in genetics, in botany, in physics, in molecular motion. Therefore, it is not surprising that probability theory has very wide applications. In the next section, we will consider one such application.

4. ODDS AND FAIR BET

A major application of probability theory is figuring out odds. People, especially in gambling or business, need to determine the odds in favor of certain actions or events before they can make a decision. For example, you go to the dentist, and the dentist tells you that you need to have 'periodontal surgery' that will cost you 1000 dollars. Before you decide to undergo this surgery costing you a huge sum, not usually covered by dental insurance, it is perhaps advisable to find out the odds in favor of being cured by the surgery. If you find that 90 out of 100 people in your age group were cured of their conditions by this surgery then the odds are in favor of such an operation. If you find out that only 20 out of 100 people in your age group were cured by the surgery, then the odds are not very much in favor and you have to weigh them against the 1000 dollars.

In the same way, in a horse race you want to bet that a certain horse would win a race. Before you make your bet, you should determine the odds in favor of that horse and decide what a fair bet should be.

DEFINITION. The *odds in favor* of an event E are $\Pr[E] : \Pr[\overline{E}]$.

It follows that if $\Pr[E] = s$, then since $\Pr[\overline{E}] = 1 - s$, the odds in favor of E are

$$s : 1 - s \quad (s \text{ to } 1 - s).$$

In figuring odds, one can find $\dfrac{s}{1 - s}$, express it in its simplest form and write the ratio of the numerator to the denominator. Thus, if the probability that a horse would win is 2/5, then the odds in favor of its winning are 2:3, since

$$s = \frac{2}{5} \text{ and } 1 - s = 1 - \frac{2}{5} = \frac{3}{5} \quad \text{so that} \quad \frac{s}{1 - s} = \frac{2/5}{3/5} = \frac{2}{3}.$$

In this case, a fair bet would be to receive 3 dollars if the horse wins and pay 2 dollars if it loses. To decide whether a bet is fair, we use the concepts of expected loss and expected win. A person's

Expected loss = (Amount of loss) × (the probability of losing).

Expected win = (The amount of win) × (the probability of winning).

A bet is *fair to a person if his expected loss is equal to his expected win.*

In the above example, since the horse has a probability 2/5 of winning, your expected win is 3 × (2/5) or 6/5 dollars. Since the probability of losing is 3/5, your expected loss is 2 × (3/5) = 6/5 dollars, and since it is equal to your expected win the bet is fair. The following theorem gives a rule to obtain the probability of an event when the odds in favor of the event are known.

THEOREM 5.4. *If the odds in favor of an event E are s : t, then*

$$Pr[E] = \frac{s}{s+t} \qquad\qquad (4)$$

Proof.   We have $\dfrac{Pr[E]}{1 - Pr[E]} = \dfrac{s}{t}$

Therefore, $t Pr[E] = s(1 - Pr[E])$ or $(s + t)Pr[E] = t$

Hence, $Pr[E] = \dfrac{s}{s+t}$.

CORROLARY. *If the odds in favor of an event E are s : t, then $Pr[\overline{E}] = \dfrac{t}{s+t}$.*

DEFINITION. If the odds in favor of E are s : t, then the *odds against* E are t : s.

Examples 5.5

1.   A card is drawn at random from a deck. You wish to bet that it is an honor card. You agree to pay $3.00 if it is not an honor card and receive $6.00 if it is an honor card.

     (a)   What is your expected win?

     (b)   What is your expected loss?

     (c)   Is it a fair bet?

Solutions.   (a) Expected win = Amount you win × Pr[winning] = $6 \times \dfrac{16}{52} = \dfrac{24}{13}$

   (b) Expected loss = Amount you loss × Pr[losing] = $3 \times \dfrac{36}{52} = \dfrac{27}{13}$

   (c) No because $\dfrac{24}{13} \neq \dfrac{27}{13}$.

2.   The odds in favor of a horse winning a race are 4:7.

   (a) What is the probability that the horse will win the race?

   (b) What is the probability that the horse will not win the race?

Solution.   (a) $\dfrac{4}{4+7} = \dfrac{4}{11}$.   (b) $\dfrac{7}{4+7} = \dfrac{7}{11}$.

Exercises 5.5

1.   Find the odds in favor of each the following events:

   (a) a card selected at random from a deck of playing cards is a heart.

   (b) a coin will come up heads both the times when it is tossed twice.

   (c) a 6 turns up when a die is rolled.

   (d) two cards chosen at random from a deck are both hearts.

   (e) the sum of the two numbers that turn up is 10 when two dice are rolled once.

2.   Find the odds in favor of each of the following events:

   (a) a card selected at random from a deck of playing cards is an ace.

   (b) a coin will come up heads at least once when it is tossed twice.

   (c) a 5 turns up when a die is rolled.

   (d) two cards chosen at random from a deck are both kings.

   (e) the sum of the two numbers that turn up is 8 when two dice are rolled once.

3.   A card is drawn at random from a deck. You wish to bet that it is a heart. You agree to pay $2.00 if it is not a heart and receive $4.00 if it is a heart.

   (a) What is your expected win?

   (b) What is your expected loss?

   (c)   Is it a fair bet?

4.   A card is drawn at random from a deck. You wish to bet that it is an honor card. You agree to pay $4.00 if it is not an honor card and receive $9.00 if it is an

honor card.

(a) What is your expected win?

(b) What is your expected loss?

(c) Is it a fair bet?

5. Two coins are tossed. You will pay 1 dollar if both of them come up heads and receive 1 dollar if they do not. Is this a fair bet?

6. A die is rolled. If the number six turns up you will receive 4 dollars, and if it doesn't, you will pay 2 dollars. Is it a fair bet?

7. A horse has a 3/4 probability of winning a race. What are the odds in favor of the horse winning? What would be a fair bet?

8. A horse has a 2/5 probability of winning a race. What are the odds in favor of the horse winning? What would be a fair bet?

9. The odds in favor of a horse winning a race are 3 : 5. (a) What is the probability that the horse will win? (b) What is the probability that the horse will not win?

10. A meteorologist forecasts that the next winter is going to be very severe. The odds in favor of his forecast coming true are 4 : 1. What is the probability that it will be a severe winter next year?

11. A roulette wheel has 38 numbers, the whole numbers 0 to 36 and 00 are marked on equally spaced slots. These spaces are alternately colored red or black, except 0 and 00, which are colored green. The wheel is spun and a ball comes to rest in one of the 38 slots. Which of the following bets are fair?

(a) You put a stake on a given number. If the ball comes to rest on this number, then you get 37 times your stake (that is, if you put 1 dollar in, you get 37 dollars back). Otherwise, you lose your stake.

(b) You put a stake on 'red'. If the ball stops on a red slot, you get twice your stake. Otherwise, you lose your stake.

12. A man offers to bet dollars to nickels that the next President of the United States is going to be a woman. What must the probability of the next President being a woman be if this is to be a fair bet?

13. Two students arrange two blind dates from a sorority. From what they know of

the sorority, they figure that the odds are 1 : 1 that both girls are blondes. How many girls are in the sorority and how many of them are blondes?

[Hint: Let the number of girls be n and the number of blondes be b. then

$$\Pr[\text{both girls are blondes}] = \frac{{}_bC_2}{{}_nC_2} = \frac{b(b - 1)}{n(n - 1)} = \frac{1}{2}$$

Hence b and n are two numbers such that the ratio of b(b - 1) : n(n - 1) is 1 : 2. To determine b and n, using a calculator or otherwise, find n(n - 1) for different values of n and determine, for example $2 \times 1 = 2$, $3 \times 2 = 6$, $4 \times 3 = 12$ and so on. From these products pick out the values for n and b that would give the answer. In this case, n = 4 and b = 3 gives the answer. Find another answer.]

14.   (a) Answer problem 13 if the odds are 1:4 that both of them are blondes.

(b) Answer problem 13 if the odds are 1 : 9 that both girls will be blondes.

15.   Bush, Carter, and Ford are ready to run a race. The odds are 4 : 9 in favor of Bush and 1 : 6 in favor of Carter. What should the odds in favor be of Ford if those three are the only runners?

16.   Liz, Sophia, and Ava are ready to run a race. The odds are 1 : 4 against Liz and 22 : 3 against Sophia. What should the odds against Ava be if those three are the only runners?

5. CONDITIONAL PROBABILITY

Very often it is necessary to change the probability of an event in view of given information about the experiment. For example, consider the following problem.

15 students out of a total of 40 students in a finite class have math anxiety. There are 18 males in the class and 5 of them have math anxiety. If you select a student at random of that class, the probability that the selected student has math anxiety is $\frac{15}{40}$. But if we ask what is the probability that the student has math anxiety given that the student is male, then the answer is $\frac{5}{18}$ because the additional information that the student is male reduces

the sample space to the 18 males. This is a case of conditional probability. In conditional probability we have an event , say E, and a condition, say F. The condition is usually stated after the phrase 'given'. In our example,

E: the student has math anxiety

F: the student is male.

We are to determine the probability of E given F, which is written as Pr[E | F].

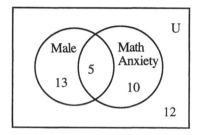

Originally we have

$$\Pr[E] = \frac{13}{40} \quad \text{and} \quad \Pr[F] = \frac{18}{40}.$$

The statement 'given F' implies that the person is male and therefore all the cases in the complement of F cannot happen, so only those outcomes of E are possible that are in E ∩ F, that is the males who experience math anxiety. Knowing the selected person is male we are restricted to only the 5 males out of 18. So,

$$\Pr[E \mid F] = \text{Probability of E given F} = \frac{5}{18}.$$

Conditional probability provides a method of obtaining the probability of an event when additional facts are known. With the help of conditional probability, it is possible to determine whether two events are independent or not. We now give the general definition.

DEFINITION. Let E and F be two events in the same sample space with $\Pr[F] \neq 0$. Then the *probability of E given F*, denoted by Pr[E | F], is given by the following:

$$\Pr[E \mid F] \quad = \quad \frac{\Pr[E \cap F]}{\Pr[F]} \tag{5}$$

Two events E and F are *independent* if $\Pr[E \mid F] = \Pr[E]$.

THEOREM 5.5. *Two events E and F, in the same sample space, are independent if and only if*

$$Pr[E \cap F] = Pr[E] \times Pr[F] \qquad (6)$$

Proof. It follows that if (6) is satisfied then (5) is satisfied and conversely.

We can illustrate Theorem 5.5 by considering equiprobable measure. At first we have

$$Pr[E] = \frac{n(E)}{n(U)} \qquad Pr[F] = \frac{n(F)}{n(U)}$$

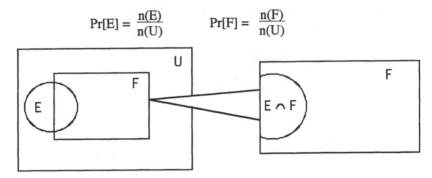

Figure 5.4

The statement 'given F' implies that F has happened and therefore all the cases in the complement of F cannot happen, so the only outcomes of E that are possible are the ones in E ∩ F. Hence the probability of E given F is n(E ∩ F)/n(F), which may be written as

$$Pr[E \mid F] = \frac{n(E \cap F)}{n[F]} = \frac{n(E \cap F)/n(U)}{n(F)/n(U)} = \frac{Pr[E \cap F]}{Pr[F]} \cdot$$

THEOREM 5.6. *Let E and F be two events in the same sample space with equiprobable measure, then*

$$Pr[E\mid F] = \frac{n[E \cap F]}{n(F)} \qquad (7)$$

Note 1. This theorem should be applied if and only if equiprobable measure is used.

Note 2. Conditional probability involves dividing one probability by another, and since a probability is usually a fraction, one must know the rules of dividing a fraction by a fraction. Note the following rule:

$$\frac{a/b}{c/d} = \frac{a}{b} \times \frac{d}{c} \cdot$$

Therefore, $$\frac{2/3}{5/6} = \frac{2}{3} \times \frac{6}{5} = \frac{4}{5}.$$

Examples 5.6

1.   A card is drawn at random from a deck of cards. What is the probability that the card drawn is a heart given that it is an honor card?

Solution.   Here E: 'the card drwn is a heart,'   F: 'it is an honor card.' So,

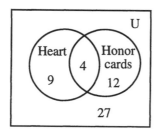 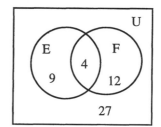

$$Pr[E] = \frac{13}{52} = \frac{1}{4}, \qquad Pr[F] = \frac{16}{52} \quad \text{and } Pr[E \mid F] = \frac{n(E \cap F)}{n(F)} = \frac{4}{16} = \frac{1}{4}.$$

E and F are independent since $Pr[E \mid F] = Pr[E]$.

2.   Find each of the given probability using the Venn diagram below.
    (a) $Pr[E]$ (b) $Pr[E \cap F]$ (c) $Pr[E \mid F]$ (d) $Pr[F \mid E]$

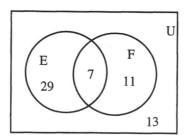

Solution. From the Venn diagram, we get $n(U) = 60$, $n(E) = 36$, $n(F) = 18$ and $n(E \cap F) = 7$.   So,

(a) $Pr[E] = \frac{36}{60}$ (b) $Pr[E \cap F] = \frac{7}{60}$

(c) $Pr[E \mid F] = \frac{7}{18}$ (d) $Pr[F \mid E] = \frac{n(F \cap E)}{n(E)} = \frac{7}{36}.$

3.   Two dice are rolled once. What is the probability that the the sum of the two numbers that come up is 10 given that at least one of them comes up 4 ?

Solution. Let   E: 'the sum of the two numbers that come up is 10'

                 F: 'at least one of them comes up 4.'

$n(U) = 6 \times 6 = 36$

$E = \{(4,6), (6,4), (5,5)\}$

$F = \{(4,1), (4,2), (4,3), (4,4), (4,5), (4,6), (6,4), (5,4), (3,4), (2,4), (1,4)\}$

$E \cap F = \{(4,6), (6,4)\}$

Therefore, $n(E) = 3, n(F) = 11, \; n(E \cap F) = 2 \;$ and

$$\Pr[E \mid F] = \frac{n[E \cap F]}{n[F]} = \frac{2}{11} .$$

E and F are not independent because $\Pr[E \mid F] \neq \Pr[E]$.

4.   A student takes a three question true/false test. What is the probability that
     E: 'he will answer all the questions correctly by guessing?'
     G: 'he will answer all the questions correctly by guessing if it is given that the teacher puts more true questions than false questions?'

Solution. Since each question may be answered in 2 ways, there are $2 \times 2 \times 2 = 8$ possible answers, one of which is the correct answer. Hence $\Pr[E] = \dfrac{n(E)}{n(U)} = \dfrac{1}{8}.$

The event corresponding to G may be reduced to a conditional probability as $\Pr[G] = \Pr[E|K]$, where K: the teacher puts more true than false questions.

$K = \{TTT, TTF, FTT, TFT\}$, so, $\Pr[K] = 4/8$.

$\Pr[E \cap K] = 1/8$ because only one element of K is the correct answer. Thus,

$$\Pr[G] = \Pr[E \mid K] = \frac{\Pr[E \cap K]}{\Pr[K]} = \frac{1/8}{4/8} = \frac{1}{4} .$$

Since $\Pr[E \mid F] \neq \Pr[E]$, the two events are not independent.

5.   Two cards are drawn one after the other from a deck of 52 cards, without replacement. What is the probability that both cards are 9's given that the first one is a 9?

Solution.   This is an experiment with equiprobable measure. There are 52 possibilities for the first card and with each of these there are 51 possibilities for the second, so $n(U) = 52 \times 51$. Let

E = both cards are 9

F = the first one is a 9. Then

$n(E) = 4 \times 3$

$n(F) = 4 \times 51$

$n(E \cap F) = 4 \times 3$

Therefore, $Pr[E \mid F] = \dfrac{n[E \cap F]}{n[F]} = \dfrac{4 \times 3}{4 \times 51} = \dfrac{3}{51}$.

6.   A letter is chosen from the word 'mathematics'. What is the probability that it is a vowel given that it is a letter in the word 'matic'

Solution. Since equiprobable measure cannot be used in this case, we cannot use theorem 5.6. Instead we use formula (5). Thus,

$Pr[E \cap F] = 3/11$, $Pr[F] = 8/11$, and

$Pr[E \mid F] = \dfrac{3/11}{8/11} = \dfrac{3}{11} \times \dfrac{11}{8} = \dfrac{3}{8}$.

7.   One card is drawn from a deck of cards, has its color noted, and is put back. A second card is then drawn. What is the probability that the second card is black given that the first card is black?

Solution.   Let   E: 'the second card is black'

F: 'the first card is black'.

$n(U) = 52 \times 52$, $n(E) = 52 \times 26$, $n(F) = 26 \times 52$, and $n(E \cap F) = 26 \times 26$.
Therefore,

$$Pr[E \mid F] = \frac{26 \times 26}{52 \times 26} = \frac{26}{52} = Pr[E]$$

Hence E and F are independent since $Pr[E \mid F] = Pr[E]$.

Exercises 5.6

1. Find each of the following using the Venn diagram.
    - (a) Pr[E]
    - (b) Pr[E ∩ F]
    - (c) Pr[E | F]
    - (d) Pr[F | E]

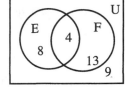

2. Find each of the following using the Venn diagram.
    - (a) Pr[E]
    - (b) Pr[E ∩ F]
    - (c) Pr[E | F]
    - (d) Pr[F | E]

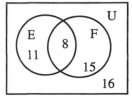

3. Find each of the following using the Venn diagram.
    - (a) Pr[F]
    - (b) Pr[E ∩ F]
    - (c) Pr[E | F]
    - (d) Pr[F | E]

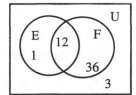

4. Find each of the following using the Venn diagram.
    - (a) Pr[F]
    - (b) Pr[E ∩ F]
    - (c) Pr[E | F]
    - (d) Pr[F | E]

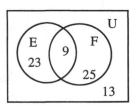

In each of problems 5 - 17 ,
    - (a) Reduce the given event, whose probability is to be determined, into two events E and F such that Pr[E | F] would give the required probability.
    - (b) Find Pr[E].
    - (c) Find Pr[E | F]   (use Venn diagrams if you like).
    - (d) Determine if E and F are independent.

5.   A card is drawn at random from a deck of cards. What is the probability that it is a spade given that it is not a face card?

6.   A card is drawn at random from a deck of cards. Find the probability that it is a heart given that it is not an honor card.

7.   A card is drawn at random from a deck of cards. What is the probability that it is a heart given that it is not a spade?

8.   A card is drawn at random from a deck of cards. What is the probability that it is a 10 given that it is a red card?

9.   A die is rolled twice. Find the probability of each of the following events:
       (a) 'the sum of the two numbers that come up is 8 given that one is a 4'.
       (b) 'the sum of the two numbers that come up is 8 given that the first is a 5'

10.   A die is rolled twice. Find the probability of each of the following events:
       (a) 'the sum of the two numbers that come up is 9 given that one is a 4'.
       (b) the sum of the two numbers that come up is 9 given that the first is a 4'.

11.   Two dice are rolled once. What is the probability that a multiple of 3 turns up given that an even number turns up?

12.   There are two birds in a cage, one green and one red.
       (a) What is the probability that both birds are male given that at least one of them is a male?
       (b) What is the probability that both birds are male given that the red bird is male?

13.   A coin is tossed three times. What is the probability that it comes up heads at least twice given that the first toss is heads?

14.   A quarter, a dime, and a nickel are tossed.
       (a) What is the probability that the nickel comes up heads given that the quarter comes up heads?
       (b) What is the probability that the nickel comes up heads given that one of the coins comes up heads?

15.   Two cards are drawn one after the other and at random from a deck of cards. Find the probability that

(a)   both cards are hearts given that the first card is a heart.

(b)   both cards are hearts given that both cards are red.

(c)   both cards are hearts given that none of the cards is a king.

16.   A card is drawn at random from a deck of cards. Its suit (diamond, club, heart, or spade) is noted and the card is put back. Then a second card is drawn. What is the probability that both cards are of the same suit given that the first is a heart?

17.   A student takes a five question true-false exam. Find the probability of each of the following events:

(a)   'he will get the correct answer if he is only guessing'.

(b)   'he will get the correct answer if he is only guessing and it is given that the instructor puts more true than false questions on his exam'.

(c)   'he will get the correct answer if it is given, in addition to (b), that the instructor never puts three questions in a row with the same answer'.

(d)   'he will get the correct answer if it is given, in addition to (b) and (c), that the first and the last question must have opposite answers'

(e)   'he will get the correct answer if it is given, in addition to (b), (c), and (d), that the answer to the second problem is F'.

18.   A student takes a ten question true-false exam. What is the probability that he will get all answers correct if

(a)   ' he is only guessing'.

(b)   ' he is only guessing and it is given that the number of questions with 'true' as the answer is the same as the number of questions with false' as the answer'

(c)   'it is given in addition to (b) that the answers to two consecutive questions are not the same'.

(d)   ' it is given, in addition to (c), that the answer to the third problem is T'.

19.   A man has three sons and gave up all hope of ever having a daughter. His wife argued that they should try once more, saying that, the fourth has to be a daughter. Criticize the validity of his wife's argument by finding the probability of the fourth child being a girl given that the first three are boys.

20.   A die is loaded in such a manner that the probability of a given face turning up is proportional to the number of dots on the face. What is the probability of rolling an odd number, given that one does not turn up?

21.   A letter is chosen at random from the word 'antidisestablishmentarianism'. What is the probability that the letter is a vowel given that it is a letter in the word 'establishment' ?

CHAPTER TEST 5

1.   Write the formula $Pr[E \cup F] = Pr[E] + Pr[F] - Pr[E \cap F]$ in words without using set notation. Then decide the advantages and the disadvantages of using set language in probability theory.

2.   A letter is chosen at random from the word 'succession'.
        (a)   What is the probability that the letter is an 's'?
        (b)   What is the probability that the letter is a vowel?

3.   A card is drawn at random from a deck of cards. Let E: It is a diamond, F: It is an honor card (ace, king, queen, or jack). Find each of the following:

        (a) $Pr[\ \overline{E}\ ]$ (b) $Pr[E \cap F]$ (c) $Pr[E \cup F]$ (d) $Pr[E \cap \ \overline{F}\ ]$

4.   A coin is tossed 4 times. Let E : 'It comes up heads exactly twice'. Find each of the following:

        (a) n(U)   (b) n(E)   (c) Pr[E]   (d) n($\overline{E}$)   (e) $Pr[\ \overline{E}\ ]$

5.   A fair coin is tossed 10 times. What is the probability that it will come up heads exactly twice?

6.   (a) Two dice are rolled once. What are the odds in favor of at least one die coming up with six?
        (b) The odds in favor of a horse winning a race are 2:3. What is the probability that the horse will win the race ?

7.   Two cards are chosen at random from a deck . You wish to bet that they are both hearts. You agree to pay $2.00 if they are not both hearts, and receive $32.00 if they are both hearts.
        (a) What is your expected win?
        (b) What is your expected loss?
        (c) Is it a fair bet?

8. Two students arrange two blind dates from a sorority. From what they know of the sorority, they figure that the odds are 2 : 1 that both girls are blondes. How many girls are in the sorority and how many of them are blonde?

9. Find each of the following using the Venn diagram.

    (a) Pr[E]

    (b) Pr[E ∩ F]

    (c) Pr[E | F]

    (d) Pr[F | E]

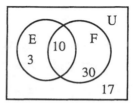

10. Two dice are rolled once. Find the probability of each of the following events:

    (a) 'the sum of the two outcomes is 10'.

    (b) 'the sum of the two outcomes is 10 given that one of them is a 4'.

11. Two cards are drawn one after the other and at random from a deck of cards. Find the probability that

    (a) 'both cards are diamonds given that the first card is a diamond'.

    (b) 'both cards are diamonds given that neither of the cards is a king'.

12. A student takes a ten question true/false exam. What is the probability that he will get all answers correct if he is guessing and

    (a) it is given that the number of questions with 'true' as the answer is the same as the number of questions with 'false' as the answer?

    (b) it is given that no two consecutive questions have the same answer ?

13. List the different sex distributions that are possible for a family of five children and find the probability of each distribution.

Chapter 6

BINOMIAL PROBABILITY

1. FINITE STOCHASTIC PROCESS

In this section we discuss the method of assigning probability measures to the outcomes of a sequence of experiments.

DEFINITION. A sequence of experiments is called a *stochastic process*. A finite stochastic process consists of a finite sequence of experiments, each having a finite number of outcomes.

To determine the probability of each outcome in a finite stochastic process, it is usually necessary to use tree diagrams in order to determine the outcomes of each experiment in the stochastic process. We illustrate the procedure with some examples.

Examples 6.1

1. A boy remembers all but the last figure of his girlfriend's telephone number and decides to choose the last figure at random in an attempt to reach her. What is the probability that he will be able to talk to her if he has the money to dial only twice?

Figure 6.1

Solution. We break this up as a sequence of two experiments. The first time he dials he either gets the correct number or the wrong number. If he gets the wrong number the first time, he will dial a second time. Then he either gets the correct number or the wrong number. We first draw a tree diagram indicating these outcomes

146

[Figure 6.1]. The probability that he gets the correct number the first time is $\frac{1}{10}$ and the probability that he gets the wrong number the first time is $\frac{9}{10}$. If he gets the wrong number the first time, the probability that he gets the correct number the second time is $\frac{1}{9}$ since he is not going to dial the digit he used the first time. The probability that he would get the correct number the second time is $\frac{1}{9}$ of $\frac{9}{10}$, that is, $\frac{1}{9} \times \frac{9}{10} = \frac{1}{10}$. Hence,

the probability that he would be able to get the correct number

$$= \text{Pr[correct number first time]} + \text{Pr[correct number second time]}$$

$$= \frac{1}{10} + \frac{1}{10} = \frac{2}{10} = \frac{1}{5}.$$

2.   A coin is tossed once and A die is rolled once. Find the probability that

(a) the coin comes up heads and the die comes up with a 2?

(b) the coin comes up heads and the die comes up with an odd numbered face?

Solution.

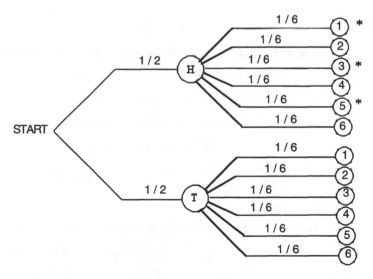

Figure 6.2

This is a finite stochastic process with two experiments. The first experiment has two possible outcomes: H or T, each with probability 1/2.

The second experiment has six possible outcomes: 1, 2, 3, 4, 5, or 6 with each outcome of the first experiment. Therefore, there are $2 \times 6 = 12$ possible outcomes for the stochastic process resulting in a tree diagram with 12 paths. Each of these 12 paths will have 2 branches, one corresponding to each of the experiments [Figure 6.2].

The probability for each branch is assigned by treating each experiment of the sequence separately. The sum of the probabilities assigned to the branches corresponding to any one of the experiments is always 1. To compute the probability of a given outcome, we find the path corresponding to the given outcome and multiply each of the probabilities occurring on the branches along the path. Thus,

$$\Pr[\text{H and 2}] = \frac{1}{2} \times \frac{1}{6} = \frac{1}{12}.$$

If a given outcome is true on more than one path, then we find the probability along each path by multiplying the probabilities on its branches. Then we add the probabilities of the paths with the given outcome.

The outcome heads and an odd number (H and odd) is true in the case of three paths (marked with *), each with probability 1/12. Hence

$$\Pr[\text{H and odd}] = 3 \times \frac{1}{12} = \frac{1}{4}.$$

The above procedure can also be applied to a finite stochastic process consisting of more than two experiments.

3.   A basket has 10 oranges, 1 apple and 1 lemon. Three fruits are taken out one after the other. Let E: 'the three fruits taken out will contain one of each kind'

   F: 'the first fruit is an apple'.

Find   (a) Pr[E]   (b) Pr[F]   (c) Pr[E ∩ F]   (d) Pr[E|F]

Solution. (See Figure 6.3)

   (a) E is true for each of the paths shown with an * . The probability for each of these paths is $\frac{10}{12} \times \frac{1}{11} \times \frac{1}{10} = \frac{1}{132}$ and since this occurs along six paths,

$$\Pr[E] = 6 \times \frac{1}{132} = \frac{1}{22}.$$

   (b) $\Pr[F] = \frac{1}{12}$

(c) $\Pr[E \cap F] = \dfrac{1}{66}$

(d) It is easy to determine conditional probabilities from a tree diagram. To determine $\Pr[E \mid F]$ assign 1 to F and 0 to any branch that does not contain an element of F. So, $\Pr[E \mid F] = \dfrac{2}{11}$.

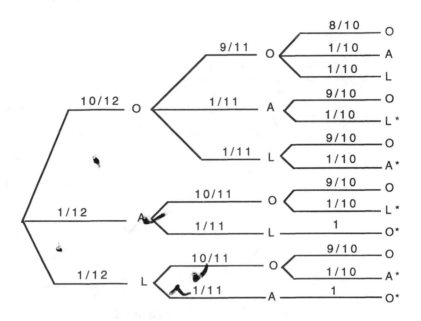

Figure 6.3

4.   In example 3, find the probability that the fruits taken out will contain one of each kind given that the first fruit is an apple.

Solution. In this case, (Figure 6.3) the first fruit is an A has a probability of 1, and the probability that the first fruit is O or L is 0. Therefore, the required probability is given by the paths AOL or ALO, that is

$$(1 \times \frac{10}{11} \times \frac{1}{10}) + (1 \times \frac{1}{11} \times \frac{10}{10}) = \frac{2}{11}.$$

5.   A die is rolled once and a coin is tossed once. What is the probability that the die comes up with 3 and the coin comes up with heads given that the die will always come up greater than 2?

Solution. This is a problem on conditional probability Pr[E|F], where

E: 3 and H,   F: greater than 2.

In this case (Figure 6.4), we assign 0 to all the outcomes not in F and assign probability to each outcome of F so that the total probability assigned is 1. Since $F = \{3, 4, 5, 6\}$, each of these are assigned a probability $\frac{1}{4}$, and 1 or 2 are assigned 0. The required probability is given by : $\frac{1}{4} \times \frac{1}{2} = \frac{1}{8}$.

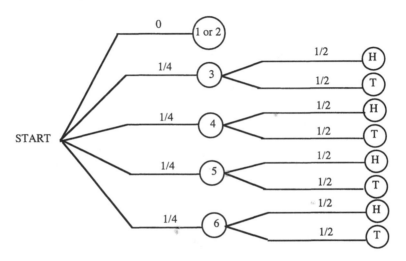

Figure 6.4

Exercises 6.1

1.   A coin is tossed once and then a die is rolled once. Find the probability that it is a head and an even number.

2.   A die is rolled once and a coin is tossed once. What is the probability that the die comes up with 1 or 6 and the coin comes up with heads?

1/6   2/6   3/6

3.   A box contains 1 white, 2 black and 3 red balls. Two balls are taken out of the box one after the other. Find the probability of each of the following:

   (a) Two black balls are taken out.

   (b) One black and one red ball are taken out.

   (c) The second ball is white given that the first ball is red.

   (d) The second ball is red given that the first ball is white.

4. A box contains 1 black, 1 white, and 3 red balls. A ball is taken out, its color is noted, then a second ball is taken out. Find the probability of each of the following:
    (a) One black and one white ball are taken out.
    (b) Two red balls are taken out.
    (c) One black and one red ball are taken out.
    (d) The second ball is red given that the first ball is white.

5. A young man was on his way to a party and forgot the directions to the house where the party was being held. All he remembered was that after taking the given exit from the highway he would come to two intersections. However, he forgot whether to go straight, take a left, or take a right on each of the intersections (it is assumed that each road at the first intersection leads to a similar intersection). Find the probability of each of the following:
    (a) He would arrive at the party by making the correct turns.
    (b) He would arrive at the party given that he made the correct turn at the first intersection.

6. A basket has 2 oranges, 1 apple, and 1 grapefruit. 3 fruits are taken out from the basket one after another. Find the probability of each of the following:
    (a) At least one orange is taken out.
    (b) One of each kind of fruit is taken out.
    (c) One fruit of each kind is taken out given that the first one is an orange.
    (d) One fruit of each kind is taken out given that the first one is an apple.

7. In a room there are two chests. Chest-1 contains two drawers, one containing one silver coin and the other containing a gold coin. Chest-2 contains two drawers, each of which contains a gold coin. You enter the room blindfolded, select a chest, open a drawer, then pick up a coin. What is the probability that the coin you pick is gold?

8. In a room there are two chests. The first chest contains two drawers, one containing two silver coins, and the other containing one gold and one silver coin. The second chest contains two drawers, one containing two gold coins and the other containing one gold and one silver coin. You enter the room blindfolded, select a chest, open a drawer, and pick up a coin. What is the probability that you would pick a gold coin?

9. A man who had a little too much to drink walks out of a bar and has to meet his wife at a motel. There are 4 roads out of the bar and he is not sure which one will take him to his motel, but he knows that it takes him 5 minutes to reach the motel. Therefore, if he does not reach the motel in 5 minutes, he will come back to the bar and try a different road. His wife told him that if he did not return by midnight, she would smash his head against the wall. If it was 11:35 PM when he got out of the bar, what is the probability that no wall in the motel was damaged?

10. A man, lost in New York city, asks the directions to the Empire State Building. The man knows that the probability of a New Yorker giving him the right directions is 0.4. The man asked two New Yorkers and both of them gave him the same directions. What is the probability that the man got the right directions to the Empire State Building?

11. A young girl is told by her father that she can wait for her boyfriend in room A or room B, but he is going to put a hungry lion in the room she does not select. The girl is shown the maze of paths (Figure 6.5) leading to the two rooms. If the boyfriend is going to choose one of the doors at random whenever he faces more than one possibility, which room should the girl select and why ? (Of course, it is assumed that the girl does NOT want her boyfriend to be devoured by a lion.)

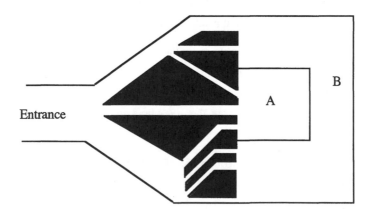

Figure 6.5

12. A young boy is told that his girlfriend has chosen to wait for him in room A or room B and there is a hungry lion in the other room. The boy is shown the maze of paths(Figure 6.6) leading to the two rooms. If the boy knows that his girlfriend has also seen the maze of paths, which room should the boy go to and why?

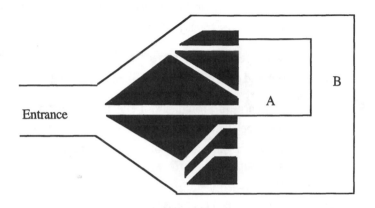

Figure 6.6

13. Two hunters shoot at a fox simultaneously. If each hunter averages one hit per three shots, what is the probability that
    (a) at least one hunter will hit the fox?
    (b) the fox will survive?

14. Two hunters shoot at a fox simultaneously. If each hunter averages one hit per four shots, what is the probability that
    (a) at least one hunter will hit the fox?
    (b) the fox will survive?

15. There are three identical boxes; box-1 contains two gold coins, box-2 contains one gold and one silver coin and box-3 contains two silver coins. One of the three boxes is chosen at random and a coin is taken out. Draw a tree diagram and indicate all the possible outcomes. What is the probability that a gold coin is taken out?

16. In problem 15, if a second coin is taken out from the box, then what is the probability that both coins are gold given that the first coin is gold?

17.   The father of a budding tennis star offers a 100 dollar prize if his son can win two sets in a row in a three-set series to be played with his father and his coach alternately: father-coach-father or coach-father-coach, according to his choice. The son is free to choose whom he plays first. The son knows that in previous encounters he had defeated his father half of the time but his coach only one third of the time. To increase his chances of winning the 100 dollar prize, whom should the son choose to play first?

[Hint: Draw a tree diagram for each of the series father-coach-father or coach-father-coach and compare the probabilities of winning.]

18.   There are two urns in a room. Urn 1 contains 2 black balls and a red ball. Urn 2 contains two black balls and two red balls. One urn is selected at random and two balls are taken out, one after the other, without replacement. Find the probability of each of the following:

(a) Both balls are red.

(b) One ball is red and the other is black.

19.   A coin is tossed four times. Draw a tree diagram for all the possibilities and assign probability measure to each branch. Find each of the following:

(a)   The total number of paths in the tree diagram.

(b)   Whether or not each path has the same probability.

(c)   The number of branches in each path.

(d)   How many paths contain the outcome: exactly twice it is heads.

(e)   The probability that it will be heads exactly twice.

20.   A coin is tossed 5 times. Answer each of the following without drawing a tree diagram:

(a)   How many paths will there be in a tree diagram for this experiment?

(b)   What is the probability for each path?

(c)   How many branches will be on each path?

(d)   What is the probability that it will turn up heads exactly three times ?

2. INDEPENDENT TRIALS WITH TWO OUTCOMES

In this section, we consider a special kind of finite stochastic process in which each experiment of the sequence is a repetition of one experiment having a given probability of success. Let us consider the following examples:

Examples 6.2

1.   A die is rolled 7 times. What is the probability that the face with six dots will turn up exactly twice?

   Solution. (See Figure 6.7). If we consider getting a six as success (S) and not getting a six as failure (F), then

$$\Pr[S] = \frac{1}{6}, \qquad\qquad \Pr[F] = 1 - \frac{1}{6} = \frac{5}{6}$$

   Therefore, rolling a die 7 times may be regarded as a finite stochastic process with a sequence of 7 experiments, each having two possible outcomes, S and F, with probabilities 1/6 and 5/6 respectively.

   In the tree diagram for this stochastic process, there will be $2^7 = 128$ different paths, each having 7 branches. A branch corresponding to the outcome S will carry a probability measure of 1/6, and the branch corresponding to F will carry a probability measure of 5/6.

   To determine the probability of 2 successes, note that each path that corresponds to this outcome will contain 2 branches with S and $7 - 2 = 5$ branches with F, so each of these paths will have a probability of $(1/6)^2 \times (5/6)^5$. Since the 2 S's may occur on any 2 out of the 7 possible branches, it follows that there are $_7C_2$ paths showing this outcome. Hence, the required probability is

$$\Pr[2 \text{ successes}] = {}_7C_2 \times (\frac{1}{6})^2 \times (\frac{5}{6})^5 .$$

   The above is an example of a finite stochastic process known as *independent trials with two outcomes*. In general, an independent trial with two outcomes is a finite stochastic process consisting of a repetition of an experiment such that

   (1)   The experiment has only two outcomes, namely success (S) and failure (F).

   (2)   The outcome of each experiment is independent of the outcomes of any previous experiment.

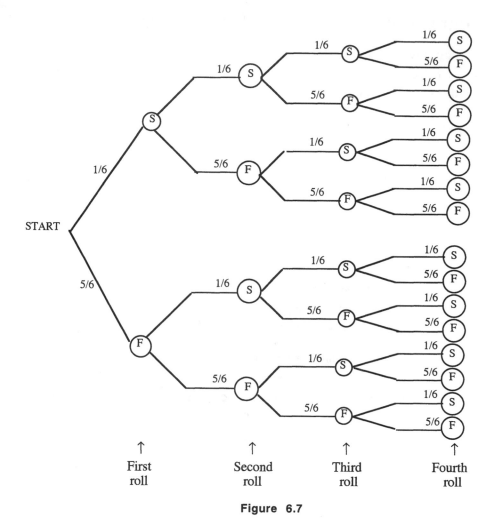

Figure 6.7

THEOREM 6.1. *Let an independent trial with two outcomes be repeated n times. If p is the probability of success, then*

$$Probability\ of\ x\ successes\ =\ (nCx)p^x(1 - p)^{n-x}.$$

Proof. Since there are 2 possible outcomes (S and F) each time the experiment is performed, the number of possible outcomes when the experiment is repeated n times is 2^n. So, the tree diagram for the resulting stochastic process will contain 2^n paths.

The required probability will be determined by the paths that have x branches with S and n - x branches with F. On each of these paths p (= Pr[S]) will appear x times and

1 - p (= Pr[F]) will appear n - x times. So, the probability on acount of each of these path is $p^x(1-p)^{n-x}$. Since there are nCx paths of this type, the required probability is nCx $p^x(1-p)^{n-x}$. Note that there are nCx paths containing x successes because x successes can occur in nCx ways.

A finite stochastic process consisting of repetitions of an independent trial with two outcomes is also called a *binomial experiment.*

Examples 6.3

1.   A die is rolled 12 times. What is the probability that it will come up sixes exactly 7 times?

Solution. In this case, $n = 12$, $x = 7$, $p = \frac{1}{6}$, and $1 - p = \frac{5}{6}$. Therefore, the required probability is $_{12}C_7 \left(\frac{1}{6}\right)^7\left(\frac{5}{6}\right)^5$.

2.   Ten persons enter an elevator at the ground floor. If there are 5 floors above the ground floor and if each person is equally likely to get off on any of the five floors, then what is the probability that exactly 4 persons will get off on the fourth floor ?

Solution. The probability that a person may get off on a particular floor is $\frac{1}{5}$ = .2. Since there are ten persons, we may approach this as a binomial probability with $n = 10$, $p = .2$, $1 - p = .8$ and $x = 4$. So, the answer is $_{10}C_4(.2)^4(.8)^6$.

Exercises 6.2

1.   A die is rolled 5 times. What is the probability that the face with six dots will come up exactly 3 times?

2.   A coin is tossed 6 times. What is the probability that it will come up heads exactly twice?

3.   In a binomial experiment, find the probability of each of the following:

(a)   Four successes in 7 trials with $p = 0.4$.

(b)   Seven successes in 10 trials with $p = 0.3$.

(c)   At least 5 successes in 7 trials with $p = 0.4$.

(d)   At least 8 successes in 10 trials with $p = 0.7$.

4.   In a binomial experiment, find the probability of each of the following:
     (a) Five successes in 8 trials with p = 0.7.
     (b) Eight successes in 12 trials with p = 0.5.
     (c)   At least 9 successes in 10 trials with p = 0.3.
     (d) At least 11 successes in 12 trials with p = 0.4.

5.   A person knows that if he plants a cherry tree he has 1/4 probability of the tree growing. What is the probability that at least one plant will grow if he plants 5 cherry trees?

6.   A doctor knows that a certain surgery he performs has 7/10 probability of success. What is the probability that he will have 10 successes if he performs the surgery 15 times ?

7.   Sixty percent of the residents of a city are democrats. What is the probability that a group of 10 people selected at random from that city will contain exactly 8 democrats?

8.   Seventy percent of the residents of a city are democrats. What is the probability that a group of 10 people selected at random from that city will contain exactly 6 democrats?

9.   To pass a ten question T-F examination, a student must answer at least 7 questions correctly. What is the probability that the student will pass if he is only guessing ?

10.   To get an 'A' in a ten question T-F examination, a student must answer at least 9 questions correctly. What is the probability that the student will get an 'A' if he is only guessing ?

11.   A coin is tossed 12 times. What is the probability that heads will occur (a) fewer than 4 times,   (b) at least 4 times?

12.   A coin is tossed 15 times. What is the probability that heads will occur (a) fewer than 2 times,   (b) at least 2 times?

13.   A card is drawn at random from a deck of cards. What is the probability that it is a face card? If this is done 7 times, what is the probability of getting a face card at least 5 times ?

14. A card is drawn at random from a deck of cards. What is the probability that it is an honor card? If this is done 10 times, what is the probability of getting a face card at least 8 times ?

15. Twelve persons enter an elevator at the ground floor. If there are 5 floors above the ground floor and if each person is equally likely to get off on any of these five floors. Find the probability of each of the following:

    $n = 12$

    $p = 1/5$ or $4/5$

    (a) Exactly 3 persons will get off on the fourth floor.
    (b) No one will get off on the fourth floor.
    (c) At least one person will get off on the fourth floor.
    (d) The elevator will stop at the fourth floor.
    (e) The elevator will stop at the second floor.
    (f) The elevator will not stop at the third floor.

16. Do problem 10 for 15 people.

17. Ten hunters shoot at a fox simultaneously. If each hunter averages one hit per three shots, what is the probability that the fox will be hit by at least one hunter ?

18. In the old days, the fortune tellers of Asia Minor used to throw astragali to predict good and evil events. An *astragalus* is the heel bone of an animal, with two broad sides and two narrower sides. The broadest side was usually labeled 4, the side opposite to 4 is 3, and the other two sides were labeled 1 and 6. When an astragalus is thrown, the probabilities of a given numbered face turning up was $Pr[4] = .39$, $Pr[3] = .37$, $Pr[1] = .12$, $Pr[6] = .12$. If an astragalus is thrown five times what is the probability that

    (a) The face 4 will appear exactly three times?
    (b) The face 6 will appear exactly twice?
    (c) The face 4 will appear the first three times and the face 3 will appear the other two times?
    (d) The face 4 will appear three times and the face labeled 1 will appear the other two times?

3. TABLE FOR BINOMIAL PROBABILITIES

It is not very easy to compute the binomial probabilities from the formula without the help of a computer or a table. Figure 6.8 is a sample of a table for binomial probabilities. Such tables are available for different values of n and different values of p. Figure 6.8 is for n = 10 and p = .1, .2, .25, .3, .4, .5, .6, .7, .8, and .9. In this table the values on the column marked x are the possible values for the number of successes. If you wish to determine the probability, say of 7 successes in 10 repetitions when   p = .6, you will look at the number 7 on the x column and then go horizontally across to the number under .6, which is .2150. Thus $_{10}C_7(.6)^7(.4)^3 = .2150$.

Pr[x successes] when
p is:

x is↓	0.1	0.2	0.25	0.3	0.4	0.5	0.6	0.7	0.8	0.9
0	0.3487	0.1074	0.0563	0.0282	0.0060	0.0010	0.0001	0.0000	0.0000	0.0000
1	0.3874	0.2684	0.1877	0.1211	0.0403	0.0098	0.0016	0.0001	0.0000	0.0000
2	0.1937	0.3020	0.2816	0.2335	0.1209	0.0439	0.0106	0.0014	0.0001	0.0000
3	0.0574	0.2013	0.2503	0.2668	0.2150	0.1172	0.0425	0.0090	0.0008	0.0000
4	0.0112	0.0881	0.1460	0.2001	0.2508	0.2051	0.1115	0.0368	0.0055	0.0001
5	0.0015	0.0264	0.0584	0.1029	0.2007	0.2461	0.2007	0.1029	0.0264	0.0015
6	0.0001	0.0055	0.0162	0.0368	0.1115	0.2051	0.2508	0.2001	0.0881	0.0012
7	0.0000	0.0008	0.0031	0.0090	0.0425	0.1172	0.2150	0.2668	0.2013	0.0574
8	0.0000	0.0001	0.0004	0.0014	0.0106	0.0439	0.1209	0.2335	0.3020	0.1937
9	0.0000	0.0000	0.0000	0.0001	0.0016	0.0098	0.0403	0.1211	0.2684	0.3874
10	0.0000	0.0000	0.0000	0.0001	0.0001	0.0010	0.0060	0.0282	0.1074	0.3487

Figure 6.8. Binomial distribution for n = 10 and various values of p.

It should be noted that $_{10}C_3(.4)^3(.6)^7$ is also .2150. In other words, the probability of 7 successes in 10 trials with p = .6 is the same as the probability of (10 - 7) = 3 successes in 10 trials with p = 1 - .6 = .4 . This is because

$$_nC_xp^x(1-p)^{n-x} = {_nC_{n-x}}(1-p)^{n-x}p^x \quad \text{since} \quad _nC_x = {_nC_{n-x}}.$$

Therefore, many binomial tables do not include $p > .5$ because those cases may be figured out from tables for p less than or equal to .5.

$$_{10}C_8 = {_{10}C_{10-8}}$$

Exercises 6.3

$n = 10$
$x = 8$
$p = .5$

1. A coin is tossed 10 times. What is the probability that
    (a) It will come up heads 8 times?
    (b) It will come up heads at least 8 times?

2. It is known that 60 percent of the residents of a city are democrats. What is the probability that a sample of 10 persons selected at random from the city will contain
    (a) 7 democrats?   (b)   at least 7 democrats?

3. A student takes a 10 question T-F test. If he is guessing, what is the probability that he will get

    $n = 10$ $_{10}C_2(.5)^2(.5)^8$
    $x = 2$
    $p = .5$

    (a) Only 2 answers correct?
    (b) At most 2 answers correct?
    (c) At least one answer correct?
    (d) At least 8 answers correct?

4. 10 persons enter an elevator at the ground floor. If there are 4 floors above the ground floor and if each person is equally likely to get off on any of these 4 floors, find the probability of each of the following:

    $n = 10$

    (a) Exactly 3 persons will get off on the second floor.
    (b) Nobody will get off on the second floor.
    (c) The elevator will stop at the second floor.
    (d) The elevator will not stop on the third floor.

    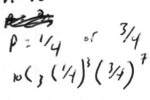
    $p = 1/4 \quad \text{or} \quad 3/4$
    $_{10}C_3(1/4)^3(3/4)^7$

5. Bolts made by a certain machine are 10% defective. In a sample of 10 bolts made by that machine, find the probability that at least 2 are defective.

6. (a) A binomial experiment with probability of success $p = .3$ is repeated 10 times. What is the number of successes that has the highest probability?
    (b) A binomial experiment with probability of success $p = .5$ is repeated 10 times. What is the number of successes that has the highest probability?

4. EXPECTED NUMBER OF SUCCESSES

If a coin is tossed 5 times, then it is possible that it may come up heads 0, 1, 2, 3, 4, or 5 times. The question is, which of these numbers should we take as the expected number of heads to occur?

Clearly, we should disregard 0 and 5 as the most unlikely, but should we expect the number of heads to be 2, 3, or 4?

In this section, we wish to develop a mathematical theory to decide what our 'expectation' should be. First, we must set our 'expectation' in a form such that it may be assigned various numerical values. Then, we should be able to compute the probability of each numerical expectation. In the given example, let's say that our expectation is: 'The number of heads in 5 tosses of a coin'. This can take any value from 0, 1, 2, 3, 4, or 5. We write

$$x = \text{possible number of heads in 5 tosses of a coin.}$$

Thus, x can be 0, 1, 2, 3, 4, or 5. We compute the probability of each of these values and make a table as shown in Figure 6.9. In the left hand column of the table, we list each possible value for x, in the central column we list the corresponding outcome, and in the right hand column we enter the probability for each of the outcomes.

x	Outcome	Probability		
0	H in no toss	$_5C_0(1/2)^0(1/2)^5$	=	1/32
1	H in any 1 of the 5 tosses	$_5C_1(1/2)^1(1/2)^4$	=	5/32
2	H in any 2 of the 5 tosses	$_5C_2(1/2)^2(1/2)^3$	=	10/32
3	H in any 3 of the 5 tosses	$_5C_3(1/2)^3(1/2)^2$	=	10/32
4	H in any 4 of the 5 tosses	$_5C_4(1/2)^4(1/2)^1$	=	5/32
5	H in all 5 of the 5 tosses	$_5C_5(1/2)^5(1/2)^0$	=	1/32

Figure 6.9

Thus, in 5 tosses of a coin, we can expect to have

0 heads with probability of 1/32

1 head with probability of 5/32

2 heads with probability of 10/32

3 heads with probability of 10/32

4 heads with probability of 5/32

and 5 heads with probability of 1/32

Then, to find the number of heads, we multiply each possible value for the number of heads by its probability and add. Thus,

The expected number of heads in five tosses of a coin

$$= 0 \times \frac{1}{32} + 1 \times \frac{5}{32} + 2 \times \frac{10}{32} + 3 \times \frac{10}{32} + 4 \times \frac{5}{32} + 5 \times \frac{1}{32}$$

$$= \frac{1}{32} \times (0 + 1 \times 5 + 2 \times 10 + 3 \times 10 + 4 \times 5 + 5 \times 1)$$

$$= \frac{80}{32} = 2.5$$

In general, the expected number of heads in n tosses of a coin

$$= (1/2^n) [0 + 1(nC_1) + 2(nC_2) + 3(nC_3) + \ldots + n(nCn)]$$

$$= (1/2^n) \times [n(1 + {}_{n-1}C_1 + {}_{n-1}C_2 + \ldots + {}_{n-1}C_{n-1})]$$

$$= (1/2^n) \times [n \times (1 + 1)^{n-1}] \qquad \textit{see problem 7(a) of exercises 11.2}$$

$$= (1/2^n) \times n(2^{n-1}) = \frac{n}{2}$$

This method may be applied to define and to find the expected number of successes for any binomial experiment consisting of a finite number of repetitions.

DEFINITION. In n repetitions of an independent trial with two outcomes, the *expected number of successes* is defined to be the sum of the products of each possible number of successes and its probability.

Thus, if n is the number of times an independent trial is repeated and p is the probability for success, then the possible number of successes x can be any of the numbers 0, 1, 2, 3, ... or n. The probability for each of these values may be tabulated as shown in Figure 6.10. The probability distribution of x for a binomial experiment is called a *binomial distribution.*

x	Probability
0	$nC_0 \, p^0 \, (1 - p)^n$
1	$nC_1 \, p \, (1 - p)^{n-1}$
2	$nC_2 \, p^2 \, (1 - p)^{n-2}$
.	. . . .
.	. . . .
n	$_nC_n \, p^n \, (1 - p)^0$

Figure 6.10

Expected number of successes

$$= 0 \times (\ ) + 1 \times (_nC_1) \, p(1- p)^{n-1} + 2 \times (_nC_2) \, p^2(1- p)^{n-2} + \ \ldots \ + n \times (_nC_n) \, p^n(1- p)^0$$

$$= np \times [(1- p)^{n-1} + (_{n-1}C_1) \, p(1- p)^{n- 2} + \ \ldots + (_{n-1}C_{n-1}) \, p^{n-1} \, (1- p)^0]$$

$$= np \times [(1- p + p)^{n-1}] = np. \quad \textit{see binomial theorem}$$

Therefore, we have the following:

THEOREM 6.2. *In n repetitions of an independent trial with two outcomes and with probability p for success, the expected number of successes is np.*

Exercises 6.4

Problems 1- 4 deal with independent trials with two outcomes. For each problem,

(a) Tabulate each possible number of successes and their probabilities.

(b) Find the expected number of successes.

1.   A coin is tossed 4 times and getting a head is regarded as a success.

2.   A coin is tossed 5 times and getting a head is regarded as a success.

3. A die is rolled 3 times and getting a six is regarded as a success.

4. A die is rolled 4 times and getting a six is regarded as a success.

5. There are 34 students in a Finite Math class. Each student shuffles a deck of cards and draws two cards at random.
    (a) How many students are likely to draw two hearts?
    (b) How many students are likely to draw two cards of the same suit?

6. There are 100 students in a Finite Math class. Each student shuffles a deck of cards and draws two cards at random.
    (a) How many students are likely to draw two honor cards?
    (b) How many students are likely to draw two face cards?

7. You plant 10 pea seeds and you know that the probability that a pea seed will germinate is 0.8. What is the expected number of seeds that will germinate?

8. A doctor knows that a certain surgery that he performs has 0.73 probability of success. If he performs the surgery on 100 patients, what is the expected number of successes?

9. The answer to each question of a 10-question multiple choice test is a, b, or c. How many questions can a student expect to answer correctly if he is only guessing?

10. A die is loaded such that a face with an even number of dots has twice as much chance of coming up as a face with an odd number of dots. In 12 throws, what is the expected number of sixes?

11. It is known that 55 percent of the residents of a city are democrats. In a random sample of 2,000 people taken from that city, what is the expected number of democrats?

12. It is known that 8 percent of those receiving Ph.D.'s in mathematics these days are women. If a newly established college is to hire 25 new Ph.D.'s (in Math) as math instructors, what is the expected number of women to be hired as math instructors?

13. At the ground floor of a hotel, 16 people enter an elevator. If each person is equally likely to get off on any of the 10 floors above the ground floor, what is the expected number of stops that the elevator makes before all 16 people get off?

5. STANDARD DEVIATION OF THE BINOMIAL DISTRIBUTION

In a binomial distribution,

(1) The number of successes with the highest probability is the expected value np. Therefore, the expected value np is regarded as the *mean* of the binomial distribution and is denoted by μ. So, in a binomial experiment

$$\text{Mean} = \mu = np .$$

(2) A number closer to the mean has a higher probability than that of a number farther away from the mean.

(3) To find the variance of the number of successes, take the square of the difference of each possible number of successes from the mean and multiply it by the corresponding probability.

That is, find $(x - \mu)^2 \times \Pr[x \text{ successes}]$, where $\mu = np$ is the mean. This number equals $np(1 - p)$. So,

$$\begin{aligned}\text{Variance } &= \Sigma(x - \mu)^2 \times \Pr[x \text{ successes}], \text{ for all possible values of x} \\ &= np(1 - p)\end{aligned}$$

(4) The square root of the variance gives the standard deviation for the number of successes. So, in a binomial experiment consisting of n repetitions of an independent trial with two outcomes and with probability p for success, the *standard deviation* σ is given by

$$\sigma = \sqrt{np(1 - p)} .$$

The number $\sqrt{np(1 - p)}$ is the standard unit of measure for standard deviation (s.d.) of a binomial experiment. Thus,

$$1 \text{ s.d.} = \sqrt{np(1 - p)}, \qquad 2 \text{ s.d.} = 2 \times \sqrt{np(1 - p)}, \qquad \text{and so on.}$$

In general,

$$k \text{ s.d.} = k \times \sqrt{np(1 - p)}$$

To find the amount of deviation when x successes occur, find the difference between x and the mean and divide it by $\sqrt{np(1 - p)}$. Thus,

$$\boxed{\text{Deviation if x successes occur} = \frac{x - np}{\sqrt{np(1 - p)}} \text{ s.d.}}$$

Examples 6.4

1. An independent trial with two outcomes and with 0.4 probability of success is repeated 600 times. If successes occur 278 times, what is the deviation in standard units?

Solution. Here n = 600, p = 0.4, and x = 278. Therefore,

$$Mean = np = 600 \times 0.4 = 240$$

$$\sigma = \sqrt{np(1 - p)} = \sqrt{600 \times 0.4 \times 0.6} = 12$$

$$Deviation\ in\ 278\ successes = \frac{(278 - 240)}{12} = 3.16\ s.d.$$

2. A coin is tossed 1000 times and it comes up heads 400 times. Give the deviation in standard units.

Solution. Here n = 1000, p = 0.5, and x = 400. Therefore,

$$Mean = 1000 \times 0.5 = 500$$

$$\sigma = \sqrt{1000 \times 0.5 \times 0.5} = 15.81$$

$$Deviation\ in\ 400\ heads = (400 - 500) \div 15.81 = -6.325\ s.d.$$

3. In a binomial experiment with n = 210,000, p = 0.3, find the range of values that the number of successes can take so that the deviation does not exceed 1 standard deviation.

Solution.

$$Mean = 210,000 \times 0.3 = 63,000$$

$$\sigma = \sqrt{210000 \times 0.3 \times 0.7} = 210$$

Hence the number of successes should be between

$$63,000 - 210 = 62,790 \quad and \quad 63,000 + 210 = 63,210.$$

In a normal distribution a deviation of the actual number of successes from the mean by one standard deviation is rather typical (in the next chapter we will show that this is the case 68% of the time), whereas a deviation exceeding one standard deviation is not very likely (32%).

Example 6.5

1. **(Which vaccine is better?)** The existing flu vaccine is known to be effective in 85 percent of the cases where it is used. Two new vaccines are claiming to be more efficient. In tests,

n = 1000

X = 860

   (a) Vaccine A proved effective in 860 out of 1,000 people who received the vaccine. Is it likely that vaccine A is more effective then the existing vaccine?

   (b) Vaccine B proved effective in 880 out of 1,000 people who received the vaccine. Is it likely that vaccine B is more effective then the existing vaccine?

Solution.   To decide this we calculate the results of the tests with the new vaccine with what would have happened if the old vaccine was given. Thus,

$880 - 850$:
$\overline{}$
11.3

   n = 1000 (Number people taking the new vaccine

   p = 0.85 (the probability of successes of the existing vaccine)

   the expected number of successes is = 1000 $\times$ 0.85 = 850 and

   the standard deviation = $\sqrt{1000 \times .85 \times .15}$ = 11.3

   (a) The deviation in 860 successes is $\dfrac{(860 - 850)}{11.30}$ = 0.9 (approx.), which is within one standard deviation from the expected number of successes and is very likely to happen with the existing vaccine. Hence, vaccine A is not likely to be more effective than the existing vaccine.

   (b) The deviation in the case of vaccine B is about 2.66, which is more than two standard deviations from the mean of the existing vaccine. Hence, vaccine B is perhaps more effective than the existing one.

Exercises 6.5

1.   A coin is tossed 900 times. What is the standard deviation for the number of times it would come up heads?

2.   A coin is tossed 1600 times. What is the standard deviation for the number of times it would come up heads?

3. A die is rolled 4,500 times. What is the standard deviation for the number of times it would come up with a six?

4. A die is rolled 18,000 times. What is the standard deviation for the number of times it would come up with a six?

5. Find the standard deviation for a binomial experiment when
   (a) $n = 10,000$, $p = 0.8$
   (b) $n = 9,600$, $p = 0.6$

6. Find the standard deviation for a binomial experiment when
   (a) $n = 2,100$, $p = 0.7$
   (b) $n = 9,600$, $p = 0.4$

7. An independent trial with two outcomes and with 0.4 probability of success is repeated 2,400 times. If success occurs 996 times, what is the deviation in standard units?

8. Do problem 4 where successes occur 936 times.

9. A coin is tossed 10,000 times and it comes up heads 5,150 times. Give the deviation in standard units.

10. Do problem 6 when the coin comes up heads 4,750 times.

11. An existing flu vaccine is known to be effective in 80 percent of the cases where it is used. The manufacturer of a new vaccine claims that the new vaccine is more effective than the existing one. In tests, it was found that the new vaccine was effective on 730 of 900 persons who received it. Does this justify the claim of the manufacturer of the new vaccine?

12. An existing polio vaccine is known to be effective on 7 out of 10 children who use it. A new polio vaccine proves effective on 1,500 out of 2,100 children who received it. Is it likely that the new polio vaccine is more effective than the existing one?

13. An unknown coin is tossed 10,000 times and it comes up heads 5,060 times. Is it likely that the coin is defective?

14.   An unknown die is rolled 50,000 times and it comes up with the same face 8,350 times. Is it likely that the die is defective?

15.   One out of ten bolts made by a machine used to be defective before the machine broke down. After the machine was repaired, it produced 8,000 defective bolts out of 40,000. Is it likely that the machine was not repaired properly?

16.   A farmer discovers that a year after a chemical plant near his farm went into operation, 1,200 out of the 10,000 corn plants that he planted died. In previous years only ten percent of his plants used to die. Should he suspect that the chemical plant is having an adverse effect on his corn plants?

17.   [**Operation Smoky,** a 1954 nuclear test, came to public attention in 1977 when Paul Cooper, a leukemia patient and former GI who took part in it, refused to accept the Veteran's Administration's rejection of his claim to disability payments. He took the case to the newspapers and succeeded in getting benefits from the Veteran's Administration shortly before he died in February 1978.]

Dr. Glen Caldwell, of the government's Center for Disease Control in Atlanta found that of the 3,153 people who took part in Operation Smoky, 8 have suffered from leukemia and 100 have suffered from other types of cancer. The expected number of leukemia cases in a normal group of that size and age range is 3.5 and that of cancer is 98.

(a) What is the probability that a person in a normal group of 3,153 people in the US will suffer leukemia? What is the standard deviation? Is there a reason to suspect that operation Smoky caused an increase in leukemia cases?

(b)   What is the probability that a person in a normal group of 3,153 people in the US will suffer from some type of cancer other than leukemia? What is the standard deviation? Are there grounds to suspect that operation Smoky caused an increase in cancer cases?

CHAPTER   TEST 6

1. Two hunters take one shot each simultaneously at a fox. Each hunter averages 1 hit per five shots. What is the probability that the fox will be hit by at least one hunter?

2. A basket has 2 red, 1 black and 1 white balls. Two balls are taken out from the basket one after the other. What is the probability that
    (a) One red and one white ball are taken out?
    (b) One of the ball is red given that the first ball is black?

3. Of three identical boxes, box-1 contains two gold coins, box-2 contains two silver coins and box-3 contains one gold and one silver coin. One of the three boxes is chosen at random and two coins are taken out one after the other. Draw a tree diagram and indicate all the possible outcomes. What is the probability that
    (a) One gold and one silver coin are taken out?
    (b) One gold and one silver coin are taken out given that the first coin is gold?

4. A young girl is told by her father that she can wait for her boyfriend in room A or room B, but he is going to put a hungry lion in the room she does not select. The girl is shown the maze of paths (Figure 6.11) leading to the two rooms. If the boyfriend is going to choose at random one of the doors whenever he faces more than one possibility, which room should the girl select and why? (Of course, it is assumed that the girl does NOT want her boyfriend devoured by the lion.)

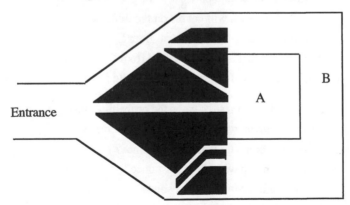

Figure 6.11

5.   (a)   A binomial experiment with probability of success p = .6 is repeated 100 times. What is the probability that success would occur exactly 60 times?

     (b)   The residents of a city are 50 percent democrats. What is the probability that a random sample of ten persons from the city will have more than 8 democrats?

6.   (a)   A binomial experiment with probability of success p = .6 is repeated 100 times. What is the number of successes that has the highest probability ?

     (b)   A binomial experiment with probability of success p = .5 is repeated 400 times. What is the number of successes that has the highest probability ?

     (c)   A binomial experiment with probability of success p = .3 is repeated 2100 times. What is the expected number of successes ?

     (d)   A die is rolled 600 times. What is the expected number of sixes?

7.   An independent trial with two outcomes and with 0.3 probability of success is repeated 2,100 times. If success occurs 600 times, what is the deviation in standard units?

8.   A student takes a ten question T-F test. To pass he must get 70 % or more answers correct. What is the probability that he would pass the test if he is only guessing?

9.   Twelve persons enter an elevator at the ground floor of a building. If there are 8 floors above the ground floor and if every person is equally likely to get off on any of these five floors then what is the probability that

     (a)   Exactly 4 persons will get off on the third floor?

     (b)   No one will get off on the third floor?

     (c)   The elevator will stop on the third floor?

     (d)   The elevator will not stop on the second floor?

10.   A die is rolled 12 times. What is the probability that

     (a)   It will come up with sixes all the twelve times?

     (b)   It will not come up with six even once?

     (c)   It will come up with six at least once?

     (d)   It will come up with six only once?

11.   When an astragalus is thrown, the probabilities of the faces turning up are: Pr[4] = .39, Pr[3] = .37, Pr[1] = .12, Pr[6] = .12. If an astragalus is thrown four times what is the probability that

      (a) The face 3 will appear the first two times and the face 6 will appear the other two times?

      (b) The face 4 will appear three times and the face 1 will appear the other time?

12.   An existing polio vaccine is known to be effective 90 percent of the time. A new polio vaccine proves to be effective on 1,450 out of 1,600 children who received it. Is it likely that the new vaccine is more effective than the existing one? Give reasons.

* LAW OF LARGE NUMBERS

Suppose we toss a coin 100 times, we expect it to come up heads $100 \times (1/2) = 50$ times. Of course, it may not come up heads exactly 50 times - it may come up heads fewer than or a little more than 50 times; say, it comes up heads 47 times. Then the ratio $\frac{x}{n}$ of the observed number of heads to the number of trials is $\frac{47}{100}$. So, the difference between the relative frequency of the number of times an outcome occurs $\frac{x}{n}$ and the theoretical probability of the outcome $(\frac{1}{2})$ is $|\frac{47}{100} - \frac{1}{2}| = .03$. In reality, there will always be a difference between $\frac{x}{n}$ and the true probability p. In other words, $|\frac{x}{n} - p|$ will in general be a non-zero number. The law of large number says that in a binomial experiment, the difference $|\frac{x}{n} - p|$ can be made as small as we need by taking n large enough. Another way of saying it is that if n is sufficiently large then the ratio $\frac{x}{n}$ of the observed number of successes to the number of trials approaches the true probability p.

Law of large numbers for a random phenomenon. If a random phenomenon is repeated a large number of times, the mean of actually observed outcomes will get closer and closer to the actual mean.

Chapter 7

BINOMIAL DISTRIBUTION AND THE NORMAL CURVE

1. BINOMIAL DISTRIBUTION

We can determine the probabilities of various outcomes of a binomial experiment by using the normal curve. Before we tell you about the normal curve and how it is used to determine probability, we need to itroduce you to the graphs of binomial distributions. Recall that we can tabulate the probability of each possible number of successes and obtain a table for the corresponding binomial distribution as was given in Figure 6.8. Now we show how to draw the graph of a binomial experiment.

Graph of a Binomial Distribution

It is possible to draw a bar graph of a binomial distribution by using the horizontal axis to represent the number of successes x and the vertical axis to represent Pr[x successes]. Thus for n = 10, we may take Pr[x] as the height of the bar for the interval [x, x + 1] and for t = 0, 1, 2, . . . , 9. In the following example (Figure 7.2), we are showing only the top of each bar because we are interested in comparing it with the normsl curve.

n = 10 and p = 0.4

x	Probability
0	0.0060
1	0.0403
2	0.1209
3	0.2150
4	0.2508
5	0.2007
6	0.1115
7	0.0425
8	0.0106
9	0.0016
10	0.0001

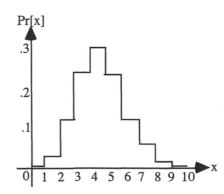

Figure 7.2

In the following example, n = 20, and we are taking Pr[x] as the height of the bar for the interval [x, x + 2] and for x = 0, 2, 4, ... , 18. Again we are showing only the tops of the bars.

n = 20 and p = 0.4

x	Probability
0	0.0000
2	0.0031
4	0.0350
6	0.1244
8	0.1797
10	0.1171
12	0.0355
14	0.0049
16	0.0003
18	0.0000
20	0.0000

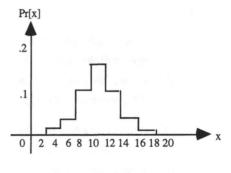

Figure 7.3

If we take a larger value of n, then the graph of a binomial distribution will drift to the right and it will appear to spread out more since the range of values for success will become larger. To see this, compare the graph in Figure 7.4 with those in Figure 7.2 and Figure 7.3.

n = 100, p = 0.4

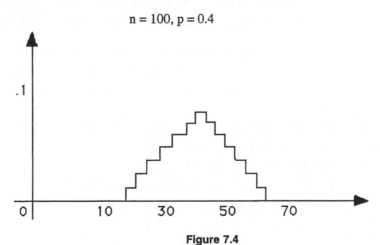

Figure 7.4

Note that the peak of the graph is usually at x = np or when x is closest to n. The total area under the graph (between the graph and the x-axis) is always 1. The graph is not, in general, symmetrical about the peak. The graph would be symmetrical about the peak if p = 0.5.

Exercises 7.1

[For these problems you may use the binomial distribution table given in Figure 6.8]

1. Draw the bar graph of the binomial distribution for n = 10 and p = 0.6.
    (a) How will the graph differ if n is increased?
    (b) What is the area under the graph?
    (c) What value of x gives the peak (highest probability)?

2. Draw the bar graph of the binomial distribution for n = 10 and p = 0.5.
    (a) How will the graph differ if n is increased?
    (b) What is the area under the graph?
    (c) What value of x gives the peak?
    (d) Is the graph symmetrical about the peak?

3. Draw the bar graph of the binomial distribution for n = 10 and p = 0.4.
    (a) What is the area under the graph?
    (b) What value of x has the highest probability?
    (c) Is the graph symmetrical about the peak?
    (d) How does this graph differ from the graph for n = 10 and p = 0.6.

4. Draw the bar graph of the binomial distribution for n = 10 and p = .7.
    (a) What value of x has the highest probability?
    (d) How does this graph differ from the graph for n = 10 and p = 0.3?

5. Draw the bar graph of the binomial distribution for n = 10 and p = 0.25.
    (a) What value of x gives the peak?
    (b) What is the area under the graph?
    (c) What value of x has the highest probability?
    (d) Is the graph symmetrical about the peak?

Z - VALUE

A binomial distribution may be approximated by the normal curve and binomial probabilities may be determined by finding suitable areas under the normal curve. To do this, it is necessary to use the standard deviation of a binomial experiment. Recall that the standard deviation of a binomial experiment with probability p and repeated n times is $\sqrt{np(1-p)}$. We rearrange the axes of the graph of the binomial experiment in the following manner.

First, to make the mean fall at 0 instead of at np, replace x by x - np.

Second, since the deviation of x - np is in terms of the standard deviation, select $\sqrt{np(1-p)}$ as the unit of measurement along the horizontal axis.

Third, multiply the unit of measurement along the vertical axis by $\sqrt{np(1-p)}$.

Consequently, the horizontal axis will give the deviation x - np in standard units. In other words, if the horizontal axis represents z, then

$$z = \frac{x - np}{\sqrt{np(1-p)}} = \frac{\text{Number of successes - mean}}{\text{Standard deviation}}$$

For any x, the number $z = [(x - np) \div \sqrt{np(1-p)}]$ is called the *z-value* of x.

The z-value of x measures the deviation of x from the mean in units of standard deviation. Thus, if the z-value of a given x is 2, then it will mean that x (actual number of successes) deviates from the mean (expected number of successes) by 2 standard deviations. Similarly, z > 2 will mean that the deviation of x is more than 2 standard deviations.

Examples 7.1

1.　In a binomial experiment, let n = 600 and p = 0.4. Find the corresponding z - value for

(a)　x = 228　　　　　(b)　x = 252　　　　　(c)　x = 294

Solution.　Mean = np = 600 × 0.4 = 240

$$\text{s.d.} = \sqrt{np(1-p)} = \sqrt{600 \times 0.4 \times 0.6} = 12.$$

(a) $z = \dfrac{228 - 240}{12} = -1$

(b) $z = \dfrac{252 - 240}{12} = 1$

(c) $z = \dfrac{294 - 240}{12} = 4.5$

Thus 228 is less than the mean by one standard deviation, 252 is larger than the mean by one standard deviation, and 294 is larger than the mean by 4.5 standard deviations.

number of successes		$z = (x - np) \div \sqrt{np(1-p)}$	
x	Pr[r successes]	z-value	(Pr[r successes]) $\times \sqrt{np(1-p)}$
0	0.0060	-2.5819889	0.00929511
1	0.0403	-1.9364917	0.06243249
2	0.1209	-1.2909944	0.18729747
3	0.2150	-0.6454972	0.3307656
4	0.2508	0	0.38853769
5	0.2007	0.6454972	0.3109231
6	0.1115	1.2909944	0.17273506
7	0.0425	1.9364917	0.06584072
8	0.0106	2.5819889	0.01642145
9	0.0016	3.2274861	0.00247871
10	0.0001	3.8729833	0.00015492

Figure 7.5. Conversion Table for n = 10 and p = .4

Since there is a z-value for each value of x, it follows that the z-value assigns a number to each outcome of the binomial experiment. If we draw the graph for a binomial distribution taking the z-value of x along the horizontal axis, and (Pr[x successes]) $\times \sqrt{np(1-p)}$ along the vertical axis, then it will have its peak at z = 0 and will spread out symmetrically about the vertical axis. This is the bar graph of a binomial distribution in normal scale. Before drawing the bar graph in normal scale, it is necessary to tabulate the z-value and (Pr[x successes]) $\times \sqrt{np(1-p)}$ for each possible value of r. We illustrate this by considering the binomial experiment for which n = 10 and p = 0.4.

For obvious reasons, one should not try to make a complete conversion table of the type given in Figure 7.5 without the help of a calculator.

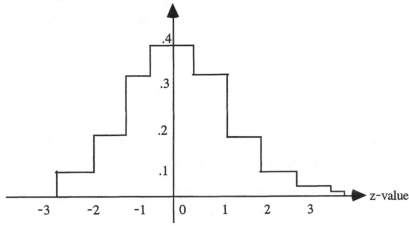

Figure 7.6. Bar graph in normal scale for n = 10 and p = .4

For the bar graph of a binomial distribution on a normal scale,

(a) The peak occurs at $z = 0$.

(b) The bar graph is symmetrical about the vertical axis.

(c) If we choose smaller and smaller intervals, the graph will closely resemble the bell-shaped normal curve.

(d) We will arrive at the same curve no matter what value of p is taken, provided we take n large enough.

Exercises 7.2

1. A binomial experiment consist of 400 repetitions of an independent trial with two outcomes and probability 0.8 for success. Find the z-value of each of the following:

    (a) 328   (b) 308   (c) 340   (d) 300

2. A binomial experiment consist of 100 repetitions of an independent trial with two outcomes and probability 0.5 for success. Find the z-value of each of the following:

    (a) 56   (b) 40   (c) 46   (d) 70

3.   A coin is tossed 144 times. It comes up heads 60 times. What is the deviation in standard units? What is the corresponding z-value?

4.   A die is rolled 180 times. It comes up six 26 times. What is the deviation in standard units of this outcome from the expected number of sixes? What is the corresponding z-value?

5.   Draw a bar graph of the binomial distribution for n = 10 and p = 0.6 after converting it to the normal scale. (Use a calculator)

6.   Draw a bar graph of the binomial distribution for n = 10 and p = 0.5 after converting it to the normal scale. (Use a calculator)

7.   A box contains 3 yellow balls and 1 white ball. A ball is taken out at random, its color noted, and put back. If this is repeated 1,200 times and a yellow ball is taken out 940 times, then what is the deviation in standard units?

2. THE NORMAL CURVE

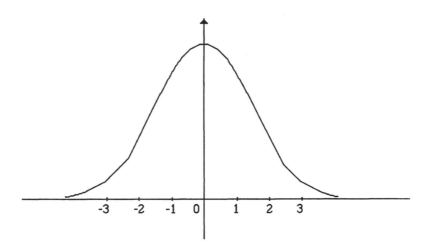

Figure 7.7. The Normal Curve

Analytically, the normal curve (Figure 7.7) is the graph of the function given by the following formula:

$$f(x) = \frac{1}{\sqrt{2\pi}} e^{-x^2/2}.$$

The Central Limit Theorem. The bar graph of a binomial distribution in the normal scale closely resembles the normal curve when n is sufficiently large.

According to the central limit theorem, the normal curve can give a good approximation of binomial probabilities when n is sufficiently large. This leads to methods of determining certain probabilities by finding the areas of certain regions under the normal curve. Therefore, it is important to learn how to find areas of given regions under the normal curve.

The total area under the normal curve corresponds to the total probability of an experiment and hence it is equal to 1. The normal curve is symmetric about the vertical axis. Therefore, the area between z = 0 and z = a is the same as that between z = 0 and z = -a.

1.   0.5 = $\left. \begin{array}{c} \text{Area under the normal curve} \\ \text{between z = 0 and z = a} \end{array} \right\}$ = $\left\{ \begin{array}{c} \text{Area under the normal curve} \\ \text{between z = 0 and z = -a} \end{array} \right.$

2.   0.5 = $\left. \begin{array}{c} \text{Area under the normal curve} \\ \text{for z} \geq 0 \end{array} \right\}$ = $\left\{ \begin{array}{c} \text{Area under the normal curve} \\ \text{for z} \leq 0 \end{array} \right.$

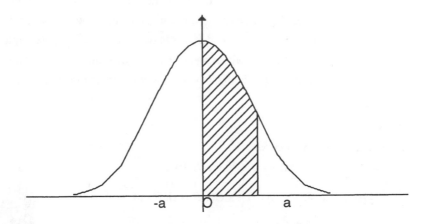

Figure 7.8. Area under the normal curve

z	A(z)
0.0	0.0000
0.1	0.0398
0.2	0.0793
0.3	0.1179
0.4	0.1554
0.5	0.1915
0.6	0.2257
0.7	0.2580
0.8	0.2881
0.9	0.3159
1.0	0.3413
1.1	0.3643
1.2	0.3849
1.3	0.4032
1.4	0.4192
1.5	0.4332
1.6	0.4452
1.7	0.4554
1.8	0.4641
1.9	0.4713

z	A(z)
2.0	0.4772
2.1	0.4821
2.2	0.4861
2.3	0.4893
2.4	0.4918
2.5	0.4938
2.6	0.4953
2.7	0.4965
2.8	0.4974
2.9	0.4981
3.0	0.49865
3.1	0.49903
3.2	0.4993129
3.3	0.4995166
3.4	0.4996631
3.5	0.4997674
3.6	0.4998409
3.7	0.4998922
3.8	0.4999277
3.9	0.4995190
4.0	0.4996830
5.0	0.4999997133

Figure 7.9. Areas under the Normal Curve

The table (Figure 7.9) for areas under the normal curve has two columns. The first column gives the positive values for z, and for each value of z, the second column under A(z) gives the area under the normal curve between 0 and z. For example, for z = 0.3, the table gives A(z) = 0.1179, which means that the area under the normal curve between z = 0 and z = 0.3 is 0.1179. We illustrate the methods of finding areas in Examples 7.2.

Examples 7.2

Area under the normal curve

1. Between z = -1 and z = 2 (Figure 7.10)

   = Area between -1 and 0 + area between 0 and 2

   = 0.3413 + 0.4772

   = 0.8185

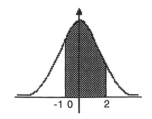

Figure 7.10

2. Between 0.8 and 2.1 (Figure 7.11)

   = Area between 0 and 2.1

      - area between 0 and 0.8

   = 0.4821 - 0.2881

   = 0.1940

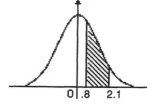

Figure 7.11

3. For z ≤ -1.6 (Figure 7.12)

   = 0.5 - area between 0 and -1.6

   = 0.5 - area between 0 and 1.6

   = 0.5 - 0.4452

   = 0.0548

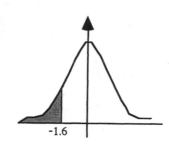

Figure 7.12

4. For z ≤ 2.7   (Figure 7.13)

   = 0.5 + area between 0 and 2.7

   = 0.5 + 0.4965

   = 0.9965

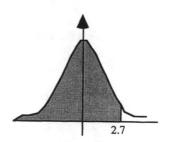

Figure 7.13

5. For z ≥ 1.3 (Figure 7.14)

   = 0.5 - area between 0 and 1.3

   = 0.5 - 0.4032

   = 0.0968

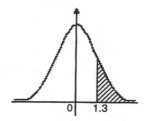

Figure 7.14

6.   Find b so that the area under the normal curve for z < b is

    (a) .89   (b) .39

Solution. (a) Since .89 > .5, it follows that b > 0 , that is, on the right of 0. Hence b should be such that the area between 0 and b is 0.89 - .5 = 0.39. From Figure 7.9, for A(z) = .39, we get z = 1.2 (approx.). Therefore, b = 1.2.

    (b) Since .39 < .5, it follows that b < 0 , that is, on the left of 0. Hence b should be such that the area between 0 and b is .5 - .39 = 0.11. From Figure 7.9, for A(z) = .11, we get z = .3 (approx.). Therefore, b = .3.

Exercises 7.3

1.   Find the area under the normal curve between
    (a) z = -1.3 and z = 2   (b) z = -1.1 and z = 3.1
    (c) z = -3 and z = -0.4   (d) z = -1.2 and z = 0
    (e) z = 0.5 and z = 2.7

2.   Find the area under the normal curve for

    (a)   z ≤ 1.7   (b) z ≤ 2.3   (c) z ≤ -1.3   (d) z ≤ -2.3

3.   Find the area under the normal curve for

    (a)   z ≥ 2   (b) z ≥ -2.1   (c) z ≥ -1   (d) z ≥ 0

4.   Find the area under the normal curve
    (a)   which is beyond z = -1 or z = 1.
    (b)   for which z ≤ -2 or z ≥ 2.
    (c)   which is beyond the interval [-1, 2].
    (d)   which is in the interval [1.1, 2.1].

5.   Find b so that the area under the normal curve for z ≤ b is

    (a) .86   (b) .16   (c) .88   (d) .115

6.   Find b so that the area under the normal curve for z ≥ b is

    (a) .9032   (b) .0548   (c) .62   (d) .0228

3. ESTIMATING BINOMIAL PROBABILITY BY THE NORMAL CURVE

We are now ready to give some practical methods of determining probability by using z-values and the central limit theorem. Let us recall that according to the central limit theorem, the bar graph of a binomial distribution in the normal scale closely resembles the normal curve when n is sufficiently large. Therefore, the normal curve can give a good approximation of binomial probabilities when n is sufficiently large. We illustrate this with some examples. The key idea is to convert a probability problem to an area under the normal curve problem by using z-values.

Examples 7.3

1.   In 1,600 tosses of a fair coin, find the probability that heads will occur more than 760 times.

Solution. $n = 1600$, $p = 0.5$, $np = 800$,   s.d. $= \sqrt{1600 \times 0.5 \times 0.5} = 20$.
It is required to find

$$\Pr[\frac{x - 800}{20} > \frac{760 - 800}{20}], \quad \text{that is,} \quad \Pr[\frac{x - 800}{20}] > -2]$$

which is equal to the area under the normal curve for $z \geq -2$. Hence, from the table for the areas under the normal curve, the answer is 0.9772.

2.   In 1,600 tosses of a fair coin, find the probability that heads will occur fewer than 830 times.

Solution. $n = 1600$, $p = 0.5$, $np = 800$,   s.d. $= \sqrt{1600 \times 0.5 \times 0.5} = 20$.
It is required to find

$$\Pr[\frac{x - 800}{20} < \frac{830 - 800}{20}], \quad \text{that is,} \quad \Pr[z \leq 1.5]$$

which is equal to the area under the normal curve for $z < 1.5$. Hence from the table for the areas under the normal curve, the answer is 0.9332.

3.   In 1600 tosses of a fair coin, find the probability that the number of heads will be between 760 and 830 times.

Solution. $n = 1{,}600$, $p = 0.5$, $np = 800$,   s.d. $= \sqrt{1600 \times 0.5 \times 0.5} = 20$. It is required to find Pr[760 ≤ x ≤ 830]. We convert the problem to one involving z-values and use corollary to the central limit theorem. Thus,

$$\Pr[\frac{760 - 800}{20} \le \frac{x - 800}{20} < \frac{830 - 800}{20}], \text{ that is,}$$

$$\Pr[-2 \le z \le 1.5] = .4772 + .4332 = .9104$$

[This is the area under the normal curve between -2 and 1.5.]

4.   In 100,000 repetitions of an independent trial with two outcomes and probability 0.8 for success, find the probability that the actual number of successes will be within 1.5 standard deviations from the mean.

Solution.   We are to find $\Pr[-1.5 < \frac{x - np}{\sqrt{np(1 - p)}} < 1.5]$. By the central limit theorem, this is equal to the area under the normal curve between -1.5 and 1.5. Hence, the answer is 0.8664.

5.   In 10,000 repetitions of an independent trial with two outcomes and with $p = 0.8$, let x be the number of successes. Find the probability that $8100 < x < 8148$.

Solution.   In this case, it will be necessary to find the deviation in standard units. Since $n = 10{,}000$ and $p = 0.8$, we get

$$np = 8000 \text{ and } \sqrt{np(1 - p)} = \sqrt{10000 \times 0.8 \times 0.2} = 40.$$

Therefore, we are to find

$$\Pr[\frac{8100 - 8000}{40} < \frac{x - 8000}{40} < \frac{8148 - 8000}{40}], \text{ that is,}$$

$$\Pr[2.5 < \frac{x - 8000}{40} < 3.7].$$

By the central limit theorem this is approximately equal to the area under the normal curve between 2.5 and 3.7, hence the answer is 0.0061.

Exercises 7.4

1. A fair coin is tossed 1,600 times. Find the probability that it will come up heads
   - (a) more than 790 times.
   - (b) fewer than 830 times.
   - (c) between 790 and 830.
   - (d) between 785 and 825.

2. In 900 tosses of a fair coin, find the probability that the number of heads will be
   - (a) more than 435 times.
   - (b) fewer than 468 times.
   - (c) between 435 and 468.
   - (d) between 440 and 470.

3. If a fair die is rolled 18,000 times, what is the probability that it will come up with a six between 2,920 and 3,020 times?

4. In 6,400 tosses of a fair coin, what is the probability that heads will occur more than 3,260 times?

5. In 900 tosses of a fair coin what is the probability that tails will occur fewer than 432 times?

6. If a fair die is rolled 18,000 times, what is the probability that it will come up with a six at least 3,130 times?

7. A true/false exam has 100 questions. Passing score is 60% or better. What is the probability that a student will pass if he is only guessing every answer?

8. A true/false exam has 100 questions. Passing score is 70% or better. What is the probability that a student will pass if he is only guessing every answer?

9. In n repetitions of an independent trial with two outcomes, find the probability that the deviation of the actual number of successes is
   - (a) between -1.5 and 1 standard units.
   - (b) fewer than -2 standard units.

10. In n repetitions of an independent trial with two outcomes, find the probability that the deviation of the actual number of successes is
    - (a) between 0.3 and 2.8 standard units.
    - (b) more than 2 standard units.

4. SOME APPLICATIONS

From the table of areas under the normal curve, we can find that

Area between -1 and 1 is 0.6826

Area beyond -1 or 1 is 0.3174

Then, from the central limit theorem, it follows that in a binomial experiment, the probability that the number of successes lie

within one standard deviation of the mean is 0.6826

beyond one standard deviation of the mean is 0.3174.

A deviation of the actual number of successes from the mean by one standard deviation is rather typical (68% of the time), whereas a deviation exceeding one standard deviation is not very likely (32%). Similarly, the probability that the number of successes will deviate from the mean by two or more standard deviations is 0.0446 (equal to the area under the normal curve beyond 2 or -2), which is a very small number. Therefore, a deviation from the mean of two or more standard deviations is surprising (4%). This simple relationship between probabilities and areas under the normal curve is of immense help in making decisions on the face of uncertainties. This is the basic principle behind the method of selecting vaccine and other problems we discussed in the last chapter. Now we demonstrate applications of these ideas to some other problems of decision.

Examples 7.4

1.   A coin is tossed 100 times. You wish to predict that it would come up heads fewer than c times and you wish to be 89% sure that your prediction is true. Determine c.

Solution. To be "89% sure that your prediction is true" means that the probability that your prediction is true is .89. Let x be the number of successes. We are to find c so that $\Pr[x \le c] = .89$. In other words, if c' is the z-value of c, then the area under the normal curve for $z \le c'$ is .89 and c' .

We first determine c'. Since .89 > .5, it follows that c' > 0 , that is, c' is on the right of 0. So, c' should be such that the area between 0 and c' is 0.89 - .5 = 0.39. (see diagram). From the table in Figure 7.9, we see that for $A(z) = .39$, we get z = 1.2 (approx.). Therefore, c' = 1.2.

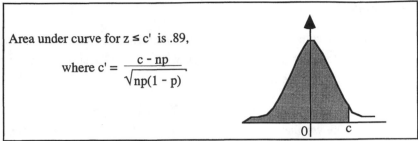

Area under curve for $z \leq c'$ is .89,

where $c' = \dfrac{c - np}{\sqrt{np(1 - p)}}$.

Then to determine c, we find $np = 100 \times .5 = 50$, $\sqrt{np(1 - p)} = 5$.

Therefore, $\dfrac{c - np}{\sqrt{np(1 - p)}} = c' = 1.2$ gives $\dfrac{c - 50}{5} = 1.2$.

Multiplying both sides by 5, we get $c - 50 = 6$, which gives $c = 56$.

So, if you predict that the coin will come up heads fewer than 56 times your prediction has 89% chance of coming true.

2.   (**How many seats?**) Four airlines are competing for 1200 passengers between New York and Niagara Falls. If a passenger is equally likely to choose any one of the airlines without regard to time and fare, how many seats should each airline provide if an airline wishes that in 89 out of 100 cases it will not have to refuse a passenger for lack of seats?

Solution.   The probability that a passenger will come to a particular airline is 0.25. Since there are 1200 passengers this is like an independent trial with two outcomes with $p = 0.25$ and $n = 1200$.

Expected number of passengers $= 1200 \times 0.25 = 300$

Standard deviation $= \sqrt{1200 \times .25 \times .75} = 15$.

We need to find a number c such that the probability that the number of passengers be fewer than c is .89. First, determine c' so that the area under the normal curve for $z < c'$ is 0.89 (See Example 1 above). Hence c should be such that the area under the normal curve between 0 and c' is $0.89 - .5 = 0.39$. This gives $c' = 1.2$. Determine c such that

$\dfrac{c - 300}{15} = 1.2$, that is $c - 300 = 1.2 \times 15 = 18$ or

$c = 300 + 18 = 318$

Hence, the number of seats to be provided is 318.

Exercises 7.5

1. A coin is tossed 400 times. You wish to predict that it would come up heads fewer than c times and you wish to be 90% sure that your prediction is true. Determine c.

2. A coin is tossed 900 times. You wish to predict that it would come up heads fewer than c times and you wish to be 90% sure that your prediction is true. Determine c.

3. A die is rolled 18,000 times. You wish to predict that it would come up with 'six' fewer than c times and you wish to be 99% sure that your prediction is true. Determine c.

4. A die is rolled 720 times. You wish to predict that it would come up six's fewer than c times and you wish to be 99% sure that your prediction will come true. Determine c.

5. A coin is tossed 900 times. You wish to predict that it would come up heads fewer than c times and you wish to be 95% sure that your prediction will come true. Determine c.

6. A die is rolled 72,000 times. You wish to predict that it would come up with 'six' fewer than c times and you wish to be 95% sure that your prediction is true. Determine c.

7. Four restaurants on Broadway are competing for 300 customers on a given Saturday after a show. If a customer is equally likely to choose any one of the restaurants at random, how many seats should each restaurant provide to be 90 percent sure that no customer will be refused because of lack of seats?

8. Two airlines are competing for 400 passengers between Providence and Hartford. If a passenger is equally likely to choose any one of the airlines without regard to time and fare how many seats should each airline provide if it wishes to be 99 percent sure that it will not have to refuse any passenger for lack of seats?

CHAPTER TEST 7

1. Draw the bar graph of the binomial distribution for $n = 10$ and $p = 0.8$.
    (a) What is the area under the graph?
    (c) What value of x has the highest probability?

2. A die is rolled 600 times. What is the probability that the deviation of the actual number of sixes from the expected number of sixes will be within 1.5 standard deviations?

3. Find the area under the normal curve
    (a) between -1.2 and 2
    (b) between 1 and 2

4. Find the area under the normal curve
    (a) for $z > -1.5$
    (b) for $z > 1.3$

5. Find b so that the area under the normal curve for $z \geq b$ is
    (a) .8849
    (b) .0445

6. A true/false exam has 100 questions. Passing score is 65% or better. What is the probability that a student will pass if he is only guessing every answer?

7. In 1600 tosses of a fair coin what is the probability that tails will occur fewer than 774 times?

8. A die is rolled 720 times. You wish to predict that it would come up six's more than c times and you wish to be 75% sure that your prediction will come true. Determine c.

9. Let a binomial experiment consist of 900 repetitions of an independent trial with two outcomes and with probability 0.8 for success. Find the probability that the number of successes will be between 702 and 726.

10. Two airlines are competing for 900 passengers between Providence and New York. If a passenger is equally likely to choose any one of the two airlines, without regard to time and fare, how many seats should each airline provide if the airline wishes to be 90% sure that it will not have to refuse a passenger for lack of seats?

Chapter 8

RANDOM VARIABLE

the mathematics of probability in action

["The greatest fault of math teaching today is the failure to teach the mathematics of gambling. I'm not talking about friendly games of chance among buddies in which all the money that goes into the pot is passed back out to those who put it in. I'm talking about cainos and lotteries in which government and private parties snitch their bit out of every pot. In those games, folks, you lose. You do. You just lose.

The more you play, the more certain it is that you will lose. It's just the mathematics of probability in action. No luck involved. Uncertainity, yes, but overall losing is as predicatable as rain.

Now, if math teachers were doing their job in educating people to this probability phenomenon, then only fools would continue to put their dollars down in support of these sneaky games. As it is, the games are stealing good people's money and destroying the lives of many.

Me, I'm an Abraham Lincoln kind of gambler. He said,"If you want to double your money, fold it over and put it in your pocket." [Wes Miller, a retired head of the mathematics department for the Warwick Public Schools, Providence Journal-Bulletin, May 10, 1996]. Let us see, if this chapter will teach you not to gamble.

1. WHAT IS A RANDOM VARIABLE

Many people buy lottery tickets almost everyday. The sale of lottery tickets soars as the prize money increases. We wish to investigate several questions. Should you buy more than one lottery ticket for the same lottery? Does your chance of winning increase when the prize money increases? Why do more people buy lottery tickets when the jackpot hits an all time high? To answer these questions, we need to understand what a random variable is.

A variable that has exactly one value corresponding to each outcome of an experiment is called a *random variable*. We usually denote a random variable by capital letters like X, Y.

Examples 8.1

1. Suppose you buy one lottery ticket for $1. The lottery offers a first prize of $10,000 and ten second prizes of $2,000 each.

Solution. In this experiment, there are three outcomes, namely, you win first prize, you win second prize or you don't win. If we say X = the number of dollars you win, then X is a random variable that will be 9999 if you win first prize, 1999 if you win a second prize, and -1 if you don't win any prize (see Figure 8.1).

2. A die is rolled. If an even number turns up, you agree to pay as many dimes as there are dots on the face that turns up, but if an odd number turns up you receive twice as many dimes as there are dots on that face.

Solution. In this case, there are six outcomes. If we say X = the number of dimes you receive, then X will be a random variable (see Figure 8.2).

DEFINITION. The *expected value* of a random variable X is the sum of the products of each possible value of X and its probability.

The concept of random variable and expected value may be used to analyze betting games. It is very similar to the concepts of expected loss and expected win which we discussed in chapter 4. In the case of a betting game, we define a random variable X as giving the amount of winnings of a particular player and treat his losses as negative winnings (see Examples 8.1). The expected value of X will then give the expected value of the game to the player. A betting game is *fair* if and only if the expected value of each player is zero. Otherwise, it is *favorable* to any player whose expected win is positive, and *unfavorable* to any player for whom the expected value is negative.

Examples 8.1

1. Suppose you buy one lottery ticket for $1. The lottery offers a first prize of $10,000 and ten second prizes of $2,000 each. If the number of tickets sold is 10,000,000,

   (a) What is the expected value of the lottery to you?

   (b) Will your expected value increase considerably if you buy 10 tickets?

   (c) What is the net gain of the lottery operation?

Solution. In this case, there are three possible outcomes: you win a first prize, you win a second prize, or you do not win any prize. In any of these outcomes, let the random variable

$$X = \text{the number of dollars you win.}$$

Then, the expected value of X will give your expected win. To determine this value, we organize the data as shown in Figure 8.1.

Outcomes	Value of X	Probability
Win first prize	9,999	$\dfrac{1}{10000000}$
Win a second prize	1,999	$10 \times \dfrac{1}{10000000}$
You do not win a prize	-1	$\dfrac{10000000 - 11}{10000000}$

Figure 8.1

$$\text{Expected win} = \frac{9999}{10000000} + \frac{1999 \times 10}{10000000} - \frac{1 \times 9999989}{10000000}$$

$$= \frac{1}{10000000} (9999 + 19990 - 9999989)$$

$$= \frac{-9970000}{10000000} = -.997$$

(a) Your expected value is -$0.997 per ticket, so, you lose $0.997 per ticket.

(b) You expect to lose ten times. That is, $9.97.

(c) The net gain is $10,000,000 × (0.997) = $9,970,000.

The lottery is unfavorable to you.

2.   A die is rolled. If an even number turns up, you agree to pay as many dimes as the number that turns up, but if an odd number turns up you receive twice as many dimes as the number that turns up. What is your expected win?

Solution. In this case, the number of dimes you receive will give a random variable X. The possible values of X and the corresponding probabilities may be tabulated as in Figure 8.2. The negative sign indicates that you have to pay.

Outcomes	Value of X	Probability
1	2	1/6
2	-2	1/6
3	6	1/6
4	-4	1/6
5	10	1/6
6	-6	1/6

Figure 8.2

The expected value of X is given by

$$2 \times \frac{1}{6} - 2 \times \frac{1}{6} + 6 \times \frac{1}{6} - 4 \times \frac{1}{6} + 10 \times \frac{1}{6} - 6 \times \frac{1}{6} = 1$$

1 dime may be regarded as the expected value of the game to you. Therefore, this game is not fair. It is favorable to you.

Expected loss appears in the expression for expected value as a negative quantity. If the expected value of a game you play is zero, then your expected win is equal to your expected loss. The terms fair, favorable, or unfavorable games should not be taken literally. The expected value of a game gives a good indication of what you may expect if you play the game a large number of times; in no way does it indicate what would happen in a single game. For example, a lottery is a very unfavorable game to each player (a person who buys one or more tickets), yet almost everybody buys lottery tickets. Of course, there is no point in buying more than one ticket for the same lottery because that does not significantly increase a person's chance of winning. Since the expected value of a lottery to a player is negative, the purchase of more than one ticket for the same lottery does not increase your chances of winning.

Exercises 8.1

1.   Suppose you buy a lottery ticket at $1. The lottery offers a first prize of $10,000, ten second prizes of $2,000 each, and 100 third prizes of $300 each. If 1,000,000 tickets are sold, answer the following:
      (a) What is the expected value of the lottery to you?
      (b) What is your expected value if you buy 100 tickets? Should you buy 100 tickets to increase your chance of winning?
      (c) What is the net gain of the lottery operators ?
      (d) Why do more people buy lottery tickets when the jackpot hits an all time high?

2.   Suppose you buy a lottery ticket for $1. The lottery offers a first prize of $1,000 and ten second prizes of $200 each. If the number of tickets sold is 10,000, answer the following:
      (a) What is the expected value of the lottery to you ?
      (b) What is your expected value if you buy 100 tickets ?
      (c) What is the net gain of the lottery operators ?

3.   A die is rolled. If an odd number turns up, you agree to pay as many dimes as the number that turns up. If an even number turns up, you receive half as many dimes as the number that turns up. Is this a fair game?

4.   A coin is tossed 4 times. You agree to pay $3 if 4 heads turn up, $1 if 3 heads turn up and $.50 if two heads turn up. In any other case, you receive $2. What is the expected value of the game to you? Is this a fair game?

5.   A chuck-a-luck cage contains two dice. A player bets $1 on any number from 1 to 6. The player is paid $2 for each die that shows his number. Is this a fair proposition?

6.   Two cards are drawn at random from a deck of cards. Suppose you agree to pay $1 if neither of the cards is an honor card (ace, king, queen, or jack), and receive $1 if both the cards are honor cards. Is this a fair game?

7.   Joe and John toss pennies. If both of them come up the same, Joe gets the pennies. Otherwise, John gets the two pennies.

   (a) What is the expected value of the game to John?

   (b) What is the expected value of the game to Joe?

   (c) John's younger brother Jim joins in the fun and says, "Every time you guys toss your pennies, I will put up one penny. If both the coins come up heads, Joe will get all three pennies; if both come up tails, John will get the three pennies; otherwise, I will get to take the three pennies." What is the expected value of the game to John? What is the expected value of the game to Jim ? Is this a fair game?

8.   A box contains one black, one green, and one white ball. Balls are successively drawn from the box without replacement until a white ball is obtained. Find the expected number of draws required. What if the box has two black, one green, and one white ball?

9.   Twenty people enter an elevator on the ground floor of a hotel that has ten floors above the ground floor. Let X be the random variable that assigns 1 if the elevator stops at the tenth floor and assigns 0 if it does not stop at the tenth floor. What is the expected value of X? What is the expected number of stops for all the twenty people to get off?

10.   A man wishes to buy an 11 cent piece of candy. In his pocket he has three dimes and three pennies. The store owner offers to let him have the candy in exchange for two coins drawn at random from the customer's pocket.

   (a) Is this a fair proposition ?

   (b) Whom would it favor if the customer had three dimes and four pennies?

11.   A chuck-a-luck cage has three dice that are shaken by turning the cage. Six players each bet $1, one on each of the six numbers. A player receives $1 for each die showing his number.

   (a) What is the expected value of the game to a player?

   (b) What is the expected value of the game to the house?

2. PROBABILITY DISTRIBUTION OF A RANDOM VARIABLE

If X is a random variable on a sample space U, then for any outcome u ∈ U,

X(u) denotes the number assigned to u by X.

The number X(u) is the *value of X at u.* For many problems, it is often helpful to make tables for u and X(u) as illustrated in Examples 8.2.

Pr[X = r] denotes the probability of X = r.

To determine Pr[X = r] add the probabilities of each outcome that is assigned the number r by X. Thus,

Pr[X = r] = The probability that the value of X is r

= The sum of the probabilities of each outcome that is assigned r by X.

A table giving r and Pr[X = r] is the probability distribution table of the random variable. In other words, the probability distribution table of a random variable gives each possible value of the random variable and the corresponding probability [See Figure 8.4].

Examples 8.2

1.   A coin is tossed 3 times. Let X be the random variable that assigns to each outcome the number of heads in the outcome. Tabulate each of the following:

(a) u, X(u)

(b) r, Pr[X = r],   and

(c) Determine the expected value of X.

Solution.   (a)

u	HHH	HHT	HTH	HTT	THH	THT	TTH	TTT
X(u)	3	2	2	1	2	1	1	0

Figure 8.3

(b) From the answer to (a), it follows that

Pr[X = 0] = 1/8,   Pr[x = 1] = 3/8,

Pr[X = 2] = 3/8,   Pr[X = 3] = 1/8.

This results in the following probability distribution:

r	0	1	2	3
Pr[X = r]	1/8	3/8	3/8	1/8

Figure 8.4

(c) In this case, the expected value of X is given by

$$0 \times \frac{1}{8} + 1 \times \frac{3}{8} + 2 \times \frac{3}{8} + 3 \times \frac{1}{8} = \frac{12}{8} = 1.5.$$

2.   A coin is tossed until a head appears or until three tails appear in a row. What is the expected number of tosses ?

Solution.   Let X give the number of tosses in each possible outcome; then the expected value of X will give the required answer. We first draw a tree diagram (Figure 8.5) for the experiment to determine the outcomes and then make a table for the probability distribution of X (Figure 8.6). To determine the expected value of X, multiply each possible value of X by its probability and then add.

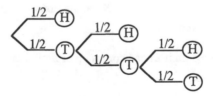

Figure 8.5

u	H	TH	TTH	TTT
X(u)	1	2	3	3

Figure 8.6

r	1	2	3
Pr[X = r]	1/2	1/4	2/8

Expected value of $X = 1 \times \frac{1}{2} + 2 \times \frac{1}{4} + 3 \times \frac{2}{8} = \frac{14}{8} = 1.75.$

The weights assigned to the elements of a sample space to determine probability measure may serve as an example of a random variable. Even though both a random variable and a probability measure may be associated with the same experiment, there are differences between them. For example, a probability measure associates a number with each subset (event) of the sample space of the experiment, whereas a random variable associates a number with each element of the sample space.

Another difference between a probability measure and a random variable is that a random variable may assign any real number, whereas the possible values of a probability measure are 0, 1, or a number between 0 and 1.

Exercises 8.2

In problems 1 - 4,

       (a) tabulate u, $r = X(u)$,

       (b) tabulate the probability distribution of the given random variable X, and

       (c) determine the expected value of X.

1.   A coin is tossed 2 times. Let X be the random variable that assigns the number of heads in an outcome.

2.   A coin is tossed 3 times. Let X be the random variable that assigns the number of tails in an outcome.

3.   A coin is tossed until a head comes up or until 4 tails in a row occur. Let X be the number of tosses.

4.   A coin is tossed until a head comes up or until 5 tails in a row occur. Let X be the number of tosses.

In problems 5 - 9, tabulate the probability distribution for the given random variable X and determine the expected value of X.

5.   A coin is tossed 3 times. You agree to pay $1 if three heads turn up and $.50 if two heads turn up. In any other case, you receive $1. Let X be the number of dollars you receive.

6.   A die is rolled. If an odd number turns up, you agree to pay as many dimes as the number that turns up. If an even number turns up, you receive half as many dimes as the number that turns up. Let X be the number of dimes you receive.

7.   A die is rolled once. Let X be the square of the number that turns up.

8.   A coin is tossed 4 times. Let X be the random variable that assigns the number of heads in an outcome.

9.   A coin is tossed 5 times. Let X be the random variable that assigns the number of heads in an outcome.

10. Two dice are rolled. Let X be the random variable that assigns to each outcome the sum of the two numbers that turn up.
     (a) What are the possible values for X?
     (b) What is $\Pr[X = 5]$?
     (c) Give the value or values of r for which $\Pr[X = r]$ is maximum.
     (d) Give the value or values of r for which $\Pr[X = r]$ is minimum.

3. MORE EXAMPLES OF RANDOM VARIABLE

Examples 8.3.

1.   A box contains 6 light bulbs of which 1 is defective. To find the defective one, the light bulbs are taken out one by one and tested. What is the expected number of light bulbs tested before the defective one is found?

Solution. Let X be the random variable that gives the number of tests. Then, the probability distribution for X is given by

r	1	2	3	4	5	6
$\Pr[X = r]$	1/6	1/6	1/6	1/6	1/6	1/6

Expected value of $X = \frac{1}{6} \times (1 + 2 + 3 + 4 + 5 + 6) = 3.5$.

2.   Two basketball players shoot the ball alternately until one of them scores or until they each miss two times. Let X be the random variable that gives the total number of shots thrown by the players. Give the probability distribution of X and the expected number of X if the probability of a success by the first player is .3 and that by the second player is .4.

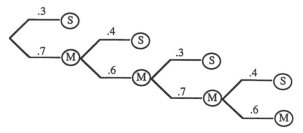

Figure 8.7

From the tree diagram in Figure 8.7 the probability distribution of X is

r	1	2	3	4
Pr[X = r]	.3	$.7 \times .4 = .28$	$.7 \times .6 \times .3 = .126$	$.7 \times .6 \times .7 = .324$

Expected value of $X = .3 + 2 \times .28 + 3 \times .126 + 4 \times .324 = 2.534$

Exercises 8.3

1. A box contains 5 light bulbs of which 1 is defective. To find the defective one, light bulbs are taken out one by one and tested. What is the expected number of light bulbs tested before the defective one is found ?

2. A box contains 8 light bulbs of which 1 is defective. To find the defective one, light bulbs are taken out one by one and tested. What is the expected number of light bulbs tested before the defective one is found ?

3. Two basketball players shoot the ball alternately until one of them scores or until they each miss three times. Let X be the random variable that gives the total number of shots thrown by the players. Give the probability distribution of X and the expected number of X if the probability of scoring by the first player is .4 and that by the second player is .6.

4. Two basketball players shoot the ball alternately until one of them scores or until they each miss three times. Let X be the random variable that gives the total number of shots thrown by the players. Give the probability distribution of X and the expected number of X if the probability of scoring by the first player is .3 and that by the second player is .6.

5.　Two cards are drawn at random from a deck of cards. What is the expected number of hearts?

6.　Two cards are drawn at random from a deck of cards. What is the expected number of face cards?

7.　A man tells his two children that in order to determine the weekly allowance each one of them would receive for the whole year, they may either toss a coin or roll a die. If the child tosses a coin and it comes up heads, then the allowance will be $20.00. If it comes up tails, then the allowance would be $ 10.00. However, if the child rolls a die, then the allowance would be as many dollars as the square of the number that turns up (that is, $ 1.00 if it is 1, $4.00 if it is 2, $9.00 if it is 3, and so on).

The children were told that each of them was free to choose either method. The son decided to toss the coin but the daughter decided to roll the die. Who made the better choice and why?

8.　Suppose your weekly allowance will be decided either by drawing a card at random from a deck or by rolling a die. If you choose to draw a card the allowance would be $35.00 if it is a face card, $4.00 if it is a spade, $3.00 if it is a club, $2.00 if it is diamond and $1.00 if it is a heart. If you roll a die the allowance in dollars would be three times the number that turns up ($ 3.00 if it is 1, $6.00 if it is 2, $9.00 if it is 3, and so on). If you are free to choose either method, which would be a better choice and why?

Expected number of Trials for the First Success

Let us now try to answer the following questions:

(a) How many times must a coin be tossed until it comes up head?

(b) How many times must a die be rolled until it comes up with 1?

(c) How many times must two dice be rolled until at least one of them comes up with a six?

One may easily guess the answer to question (a) to be 2 and that of (b) to be 6. But, it is not easy to guess the answer to (c). In this case, we use Theorem 8.1.

(a) Since $\Pr[H] = \frac{1}{2}$, the expected number of times a coin must be tossed to get

the first H is $\frac{1}{1/2} = 2$.

(b) Since $\Pr[1] \doteq \frac{1}{6}$. The expected number of times a die must be rolled to get the

first 1 is $\frac{1}{1/6} = 6$.

(c) p = the probability of getting at least one six when two dice are rolled $= \frac{11}{36}$.

So, the expected number of times two dice must be rolled so that at least one

of them comes up '6' is $\frac{1}{11/36} = \frac{36}{11} = 3.27$.

Theorem 8.1 is easy to understand and easy to apply. There is a beautiful proof of
this theorem by using the sum of a geometric series. To prove the theorem we must first
prove two lemmas [a lemma is a result that is needed to prove a theorem.]

THEOREM 8.1. *If Pr[S] = p for an independent trial, then the expected number of trials to*

get the first success is $\frac{1}{p}$.

Lemma 1. If $q < 1$, then $1 + q + q^2 + q^3 + \ldots = \frac{1}{1-q}$.

Proof of lemma 1.   Let $S = 1 + q + q^2 + q^3 + \ldots$ (i)

Multiplying both sides by q, $qS = q + q^2 + q^3 + \ldots$ (ii)

Subtracting (ii) from (i), $(1 - q)S = 1$.

Therefore, $S = \frac{1}{1-q}$.

Lemma 2. If $q < 1$, then $1 + 2q + 3q^2 + 4q^3 + \ldots = \frac{1}{(1-q)^2}$

The proof of lemma 2 is very similar to the proof of lemma 1. Let

$$T = 1 + 2q + 3q^2 + 4q^3 + \ldots$$

So $qT = q + 2q^2 + 3q^3 + 4q^4 + \ldots$

Subtracting, $(1 - q)T = 1 + q + q^2 + q^3 + \ldots$

$$= \frac{1}{1-q} \text{ by lemma 1. So,}$$

$$(1 - q)T = \frac{1}{1-q} \qquad \text{or} \quad T = \frac{1}{(1-q)^2} .$$

Proof of Theorem 8.1. Let X be the number of trials for the first success and $q = 1 - p$. Then we can tabulate the possible values of X and the respective probability for each in the following manner:

X	1	2	3	4	.	.	.
Pr[X = r]	p	pq	pq^2	pq^3	.	.	.

Expected value of $X = p + 2pq + 3pq^2 + 4pq^3 + . \quad . \qquad . \qquad .$

$\qquad\qquad\qquad = p(1 + 2q + 3q^2 + 4q^3 + \qquad . \qquad . \qquad .)$

$\qquad\qquad\qquad = p \cdot \dfrac{1}{(1 - q)^2} \quad$ by lemma 2 because $q < 1$.

$\qquad\qquad\qquad = p \cdot \dfrac{1}{p^2} \qquad$ because $1 - q = p$

$\qquad\qquad\qquad = \dfrac{1}{p} \cdot$ ∎

Exercises 8.4

1. How many times must a die be rolled until it comes with a six?

2. How many times must two dice be rolled until at least one of them comes up with one?

3. How many times must two dice be rolled until you get a sum of 5?

4. How many times must two dice be rolled until you get a sum of 7?

5. How many cards you must draw from a deck till you draw a king?

6. How many cards must you draw from a deck until you draw an honor card?

7. How many cards must you draw from a deck until you draw a face card?

8. How many cards must you draw from a deck until you draw a card with a prime number?

9. A city is 20% democrat. How many persons do you expect to meet until you meet the first democrat?

10. A city is 40% democrat. How many persons do you expect to meet until you meet the first democrat?

4. THE STANDARD DEVIATION OF A RANDOM VARIABLE

The expected value of a random variable is the *mean* of the random variable. The mean of a random variable is usually denoted by the Greek letter μ. Let X be a random variable with $x_1, x_2, \ldots, x_n$ as possible values and μ as mean (expected value). Then $x_i - \mu$ is the *deviation of* x_i. We wish to determine how closely the values of X cluster around μ. To do this, we define the *Variance* V(X) of X by the following:

$$V(X) = (x_1 - \mu)^2 Pr[X = x_1] + (x_2 - \mu)^2 Pr[X = x_2] + \ldots + (x_n - \mu)^2 Pr[X = x_n].$$

Thus, to determine V(X), we take the square of the deviation of each value from the mean and multiply it by the corresponding probability and then add.

The reason for taking the square of the deviation from the mean is to avoid negative quantities because if x_i is smaller than μ then $(x_i - \mu)$ would be negative. Since we are interested in figuring out only the deviations from the mean and not whether it is larger or smaller than the mean we take the squared quantity. However, this creates one problem : in variance the units also get squared. For example, if the unit for X is dollar then the unit for variance would become "square dollars" and if the units for X are hours then the units for V(X) will be "square hours." Since things like square hours and square dollars will not be easy to comprehend, we use the square root of the variance. The square root of the variance of a random variable X is the *standard deviation* $\sigma(X)$ of X.

$$\text{Standard Deviation of } X = \sigma(X) = \sqrt{V(X)}.$$

Examples 8.6

1. If a coin is tossed three times and the random variable X gives the number of heads, then the probability distribution of X is

r	0	1	2	3
Pr[X = r]	1/8	3/8	3/8	1/8

    So, $\mu = 1.5$ (see Examples 8.1),

    $$V(X) = (0 - 1.5)^2(\tfrac{1}{8}) + (1 - 1.5)^2(\tfrac{3}{8}) + (2 - 1.5)^2(\tfrac{2}{8}) + (3 - 1.5)^2(\tfrac{3}{8})$$

    $$= (\tfrac{1}{8})[(1.5)^2 + (-.5)^2 \times 3 + (.5)^2 \times 2 + (1.5)^2 \times 3] = \frac{12}{16}$$

    and $\qquad \sigma(X) = \dfrac{\sqrt{12}}{\sqrt{16}} = \dfrac{2\sqrt{3}}{4} = \dfrac{\sqrt{3}}{2}.$

2.   Find the mean, the variance and the standard deviation of the random variable whose probability distribution is given by

r	-5	1	2	3	4	5
Pr[X = r]	.1	.2	.2	.2	.2	.1

$$\mu = (-5) \times .1 + 1 \times .2 + 2 \times .2 + 3 \times .2 + 4 \times .2 + 5 \times .1 = 2$$

$$V(X) = (-5-2)^2 \times .1 + (1-2)^2 \times .2 + (2-2)^2 \times .2 + (3-2)^2 \times .2 + (4-2)^2 \times .2 + (5-2)^2 \times .1$$

$$= 4.9 + .2 + 0 + .2 + .8 + .9 = 7$$

$$\sigma = \sqrt{7}.$$

Exercises 8.5

1.   Determine the variance and the standard deviation for the random variable in problem 2, Examples 8.2.

2.   Determine the variance and the standard deviation of the random variable in problem 2, Examples 8.3.

3.   Two dice are rolled. Let X be the random variable that gives for each outcome the sum of the two numbers that turn up. Give the mean, the variance and the standard deviation of X .

4.   Find the mean, the variance, and the standard deviation of the random variable whose probability distribution is given by

r	35	30	50	20
Pr[X = r]	.4	.1	.1	.4

5.   Find the mean, the variance and the standard deviation of the random variable whose probability distribution is given by

r	-4	-3	-1	2	4
Pr[X = r]	.1	.2	.2	.4	.1

6.   Find the mean, the variance and the standard deviation of the random variable whose probability distribution is given by

x	-5	1	2	3	4	5
Pr[X = r]	.3	.1	.1	.1	.1	.3

7.   Three dice are rolled. Find the expected value of the sum of the three numbers that turn up and give the standard deviation.

5. CHEBYSHEV'S THEOREM

Often it is very important to determine the probability for the number of successes to be within a certain deviation from the expected number. For example, let's say you have a new car whose mpg (mileage per gallon) is 30 with a standard deviation of 2 mpg. What is the probability that with your next tankfull of gas, the mpg of your car will be between 27 and 33 ? To answer such questions, we have to use the following:

Chebyshev's Theorem. *Let X be a random variable with expected value μ and standard deviation σ. The probability that the value of X will differ from μ by less than k standard deviations is at least $1 - \dfrac{1}{k^2}$, that is,*

$$Pr[\mu - k\sigma < X < \mu + k\sigma] \geq   1 - \frac{1}{k^2} .$$

In the example of the car, we are looking for a deviation factor of

$$k = \frac{33 - 30}{2} = \frac{30 - 27}{2} = \frac{3}{2} .$$

Therefore, using this value of k in Chebyshev's theorem, we get the required probability to be $\geq \dfrac{5}{9}$ since $1 - \dfrac{1}{(3/2)^2} = 1 - \dfrac{4}{9} = \dfrac{5}{9} .$

Note 1.   The formula in Chebyshev's theorem is known as Chebyshev's inequality since it is an expression involving the inequality symbol ' $\geq$ '.

Note 2.   Chebyshev's inequality does not give anything for $k \leq 1$. If $k = 1$, then the right hand side expression in the inequality is 0 and if $k < 1$, then it is negative.

Note 3.   Chebyshev's Theorem, as given above, may be applied if either a value of k is given or if two values of X with the mean as the midpoint are given.

Note 4. To determine k when two appropriate values of X are given, apply the following formula:

$$k = \frac{\text{Higher value of X } - \text{ Mean}}{\text{Standard deviation}}$$

Note 5. Chebyshev's inequality gives a lower bound for the probability that X deviates from the mean by at most k units. That is, the value of X is within k units of its mean . By considering the complement, we get

$$\Pr[\ X \text{ is not between } \mu - k\sigma \text{ and } \mu + k\sigma] \ \leq \ \frac{1}{k^2}$$

$$\text{Or} \quad \Pr[\ X \geq \mu + k\sigma \text{ or } X \leq \mu - k\sigma] \ \leq \ \frac{1}{k^2}$$

$$\text{Or} \quad \Pr[\ |X - \mu| \geq k\sigma] \ \leq \ \frac{1}{k^2}$$

The above are alternate forms of Chebyshev's inequality and each of these forms gives an upper bound for the probability that X deviates from its mean by more than k units.

Examples 8.7

1. Let X be the random variable that gives in minutes the time you have to wait at the check-out counter of your grocery store. Let the following give the probability distribution of X based on your experience of the last 10 times you went to the grocery store. What is the probability that the next time you go to the grocery store you will have to wait between 7 and 13 minutes?

r	6	7	8	9	10	11	12	13	14
Pr[X = r]	.1	.1	.1	.1	.2	.1	.1	.1	.1

Solution. Mean of $X = (6 + 7 + 8 + 9) \times .1 + 10 \times .2 + (11 + 12 + 13 + 14) \times .1 = 10$

$V(X) = (6 - 10)^2 \times .1 + (7 - 10)^2 \times .1 + (8 - 10)^2 \times .1 + (9 - 10)^2 \times .1 + 0 \times .2$

$\qquad\qquad + (11 - 10)^2 \times .1 + (12 - 10)^2 \times .1 + (13 - 10)^2 \times .1 + (14 - 10)^2 \times .1$

$\qquad = (16 + 9 + 4 + 1) \times .1 + 0 + (1 + 4 + 9 + 16) \times .1 = 6.$

Therefore, the standard deviation σ of X is $\sqrt{6}$. Since we are looking for the probability of X to lie between 7 and 13, we get

$$k^2 = \frac{(13 - 10)^2}{(\sqrt{6})^2} = \frac{3^2}{6} = \frac{3}{2}.$$

By Chebyshev's theorem, the required probability is at least $1 - \dfrac{2}{3} = \dfrac{1}{3}.$

2.   Give the probability that the values of a random variable X be within 5/4 standard deviations of the mean. In this case, k = 5/4. Therefore, from Chebyshev's theorem, the required probability is at least

$$1 - \frac{1}{(5/4)^2} = 1 - \frac{16}{25} = \frac{9}{25}.$$

Solution.   At least $\frac{9}{25}$ $(\geq \frac{9}{25})$.

3.   A manufacturer of light bulbs claims that their lightbulbs have a mean life of 1,000 hours with a standard deviation of 50 hours.

(a) Estimate the probability that one of these light bulbs will have a life of 925 to 1075 hours.

(b) In a sample of 100 light bulbs made by that manufacturer, how many light bulbs do you estimate would have a life of 925 to 1075 hours?

Solution.   (a) $k = \frac{1075 - 1000}{50} = \frac{3}{2}$ so the probability is at least $1 - \frac{4}{9} = \frac{5}{9}.$

(b) The required number of light bulbs is at least $100 \times \frac{5}{9} = 55.5.$

Exercises 8.6

1.   For each of the following estimate the probability that the values of a random variable X will be

(a) within 2 standard deviations of the mean.

(b) within 3 standard deviations of the mean.

(c) within 5/2 standard deviations of the mean.

(d) within 3/2 standard deviations of the mean.

(e) within 4/3 standard deviations of the mean.

2.   Let X be the random variable that gives the mpg of a car calculated whenever the gas tank is filled. Let the following table give the probability distribution of X based on previous records:

r	35	30	50	20
Pr[X = r]	.4	.1	.1	.4

(a) What is the probability that the mpg of the car is between 20 and 40 miles?

(b) What is the probability that the mpg of the car is between 18 and 42 miles?

3.   A car gets 36 mpg with a standard deviation of 2 mpg. What is the probability that with the next tankful of gas the car will get between 32 and 40 mpg?

4.   A car gets 25 mpg with a standard deviation of 3 mpg. What is the probability that with the next tankfull of gas the car will get between 21 and 29 mpg?

5.   Let X be the random variable that gives in minutes the time you have to wait at the check-out counter of a grocery store. Let the following give the probability distribution of X based on your experience of the last 10 trips to the grocery store. What is the probability that the next time you go to the grocery store you will have to wait between 7 and 21 minutes?

r	5	10	15	20	25
Pr[X = r]	.2	.2	.3	.2	.1

6.   A manufacturer of light light bulbs claims that their light bulbs have a mean life of 800 hours with a standard deviation of 40 hours.
   (a) Estimate the probability that one of these light bulbs will have a life of 750 - 850 hours.
   (b) In a sample of 100 light bulbs made by that manufacturer, how many light bulbs do you estimate would have a life of 750 to 850 hours?

7.   Let X be the random variable that gives in minutes the time you have to wait at the check-out counter of your grocery store. Let the following give the probability distribution of X based on your experience of the last 10 times you went to the grocery store. What is the probability that the next time you go to the grocery store you will have to wait between 6 and 14 minutes?

r	5	6	7	8	9	10	11	12	13	14	15
Pr[X = r]	.1	.1	.1	.1	.1	0	.1	.1	.1	.1	.1

8.   A manufacturer of light bulbs claims that his light bulbs have a mean life of 1000 hours with a standard deviation of 50 hours. In a sample of 100 light bulbs made by the manufacturer, how many light bulbs do you estimate would have
   (a) a life of 960 to 1060 hours?
   (b) a life of 900 to 1100 hours?
   (c) a life of 875 to 1125 hours?

(d) a life of 850 to 1150 hours?

(e) a life of 800 to 1200 hours?

9.   The probability distribution of a random variable X has an expected value of
100 and a standard deviation of 10. What percentages of the values of X lie in
each of the following intervals?

(a) 88 - 112   (b) 85 - 115   (c) 80 - 120

(d) 75 - 125   (e) 78 - 122

[Hint: Determine the probability using Chebyshev's theorem and express it in
percentage. Thus in case of   (a) $k = 1.2$ and the probability is at least .305.
Therefore, the required answer is 30.5 percent.]

CHAPTER TEST 8

1.   (a) Give the definition and an example of a random variable.

(b) A coin is tossed 100 times. What is the expected number of heads that will
occur?

2.   Suppose you buy a lottery ticket for $1. The lottery offers a first prize of $500,
and nine second prizes of $100 each. If the number of tickets sold is 10,000, then
what is the expected value of the lottery to you? Should you buy 100 tickets to
increase your chances of winning? Give reasons.

3.   A coin is tossed 4 times. You agree to receive $1 if the number of times it
comes up heads is greater than the number of tails, otherwise you pay $1. Is this
a fair game? What if the coin is tossed 5 times?

4.   A coin is tossed until a tail comes up or until 3 heads in a row occur. What is the
expected number of tosses?

5.   A man wishes to buy an 11 cent piece of candy. He has in his pocket four dimes
and three pennies. The store owner offers to let him have the candy in exchange
for two coins drawn at random from the customer's pocket. Should the customer
agree to this game? Justify your answer.

6.   A chuck-a-luck cage contains two dice. A player bets $1 on any number from 1 to 6. The player is paid $10 if both the dice show his number, and $2 if any one of the dice show his number. What is the expected win of the house? Is this game favorable to the player?

7.   A coin is tossed 4 times. For any outcome u, let the random variable X give the number of heads in u. Find

    (a) $Pr[X = 2]$ (b) $Pr[X = 4]$

8.   Find the mean, the variance and the standard deviation of the random variable whose probability distribution is given by

x	15	12	10	9	4
$Pr[X = r]$	.3	.1	.4	.1	.1

9.   Two cards are drawn at random from a deck of cards. What is the expected number of honor cards (ace, king, queen, or jack) drawn?

10.   Let X be the random variable that gives in minutes the time you have to wait at the check-out counter of a grocery store. Let the following give the probability distribution of X based on your experience of the last 10 trips to the grocery store. What is the probability that the next time you go to the grocery store you will have to wait between 7 and 25 minutes?

r	5	10	15	20	25
$Pr[X = r]$	.1	.2	.3	.2	.2

Chebyshev's theorem holds for any kind of probability distribution that has a finite variance and it is also useful in proving various other theorems. However, for many distributions that arise in practice there are other methods that give sharper bounds for probability of deviations from the mean than the ones given by Chebyshev's theorem. For example, Chebyshev's theorem holds for binomial distribution as well. However, in case of binomial distribution the normal curve gives a more convenient and more satisfactory method of estimating probability than Chebyshev's theorem.

Chapter 9

INFERENTIAL STATISTICS

A numerical guide to decision making

So far, we have been working with the notion that the probability of an event can be determined before the event takes place. A probability computed before an outcome happens, like rolling a die, is *a priori* probability. The classical notion of probability is that of a priori probability. According to the modern notion, the existence of a probability is assumed and methods are established to estimate the probability from trials and observations. Probabilities computed after trials and observations are *posteriori* probabilities. Calculations of posterior probabilities usually require statistical methods.

The basic idea is to estimate the probability measure of an outcome by performing certain experiments and collecting some data about the outcome. For example, in an election there are two candidates, A and B. Then there are only three possible outcomes: {A wins, B wins, tie}. Does this mean that the probability of A winning is 1/3? Surely, it is not the case of an equiprobable measure. The outcome definitely depends on many factors. Perhaps the only way to get an idea would be to go and ask the voters. Then, the question is which voters to ask? If we have to ask each voter, it would be like holding a separate election. Instead, we select a sample of voters and estimate a percentage of voters in favor of each candidate as done in opinion polls.

Similarly, a meteorologist has to forecast the chances of rain on a particular day well ahead of time. On a particular day, there are only two possible outcomes: either it rains or it does not rain. Does it mean that the probability of it raining on any day is 1/2? No, since we are not dealing with equiprobable measure. Instead, the meteorologist goes through his collected data about the region and finds out how often it had rained under similar conditions (like temperature, humidity, wind velocity, etc.). On the basis of the data he or she estimates a probability measure for rain on a particular date.

This kind of estimate for probability measure involves collecting, organizing, and analyzing data. Furthermore, one needs to draw a sample and make a conclusion on the basis of information gathered from the sample.

Statistics is the mathematical science of drawing conclusions, like estimating probability measure about a population, through an analysis of data collected from a sample. *Inductive Statistics* or *Inferential Statistics* is the science of drawing conclusions

214

from the sample data. *Descriptive Statistics* deals with collecting, organizing, and analyzing data. We have already discussed Descriptive Statistics in chapter 1. In Inferential Statistics, we need to apply the techniques of Descriptive Statistics to collect and organize empirical data. Then we apply the theory of probability to make inferences. In this chapter, we shall discuss the following:

- (a) How to draw random samples.
- (b) How to estimate probability from data.
- (c) How opinion polls are conducted.
- (d) How to calculate margin of error.
- (e) Relation between sample mean and population mean.
- (e) How to estimate population size from samples.
- (f) How to use the theory of Binomial probability to determine an interval within which the true value of probability lies.

1. POPULATION AND SAMPLE

Since we need to draw conclusions about a population from observations on a sample, the terms population and sample need to be defined.

DEFINITION. The word *population* in statistics refers to any set of elements from which numerical data is collected. A *sample* is a proper subset of a population.

A statistical inference can be reliable only if the sample is selected properly. A sample is *fair* if every element of the population had the same probability of being selected. Otherwise, the sample is *biased*. In addition, for more accurate results, the sample must be large enough to correctly reflect the population. Another basic requirement of statistical inference is that the inference be drawn from random samples. Random sampling can produce fairly accurate results.

DEFINITION. If a sample is selected so that every subset of the population of the same size as the given sample is equally likely to be selected, then the given sample is a *random sample*.

Examples 9.1

1.   From a group of 50 males, a girl was asked to select a sample of 5 males. According to her own heart's desire, the girl picked 5 tall and handsome guys. Was this a fair or a biased sample?

Solution. Biased, because tall and handsome males are more likely to be selected.

2.   In order to sample food prices, a lady asked her seven year old son to pick up 10 food items from a grocery store. Her son picked up three different kinds of candy, two kinds of soda, two boxes of cereal, and three kinds of ice cream. Do these 10 items constitute a fair sample?

Solution. No, because the child seems to be interested only in what may be regarded as "kid's stuff". Hence, such items were more likely to be selected than other items.

3.   To select a random sample for the situation in example 1, assign a number to each male. Write each of these numbers on a slip of paper and fold the paper so that no one can see it without unfolding it. Then put them in a hat or a basket and mix thoroughly. The girl may be asked to draw any 5 slips of paper from the hat and the numbers on those papers would then determine the people in the sample.

4.   In example 2, a random sample may be obtained by writing down the names of all the available items in the store and drawing from the hat as in example 3. If it is impossible to do so, then the first aisles may be selected at random by writing the numbers on slips of paper and drawing from a hat. Then, the kid may be blind-folded and asked to point to any item in the aisle.

5.   How can you obtain a random sample of four cards from a deck of cards?

Solution. Shuffle the deck well and pick any four cards without looking at the faces of the cards.

6.   You wish to study the opinion of the students of this university about the benefits of college education. How can you select a random sample of students for this study?

Solution. Take a complete list of all the students from the registrar's office or the student directory. Names may then be drawn from a hat or selected by other means so that every name has the same chance of being selected.

Random Digits

Row #	Random digits										
1.	01	39	80	06	44	47	12	11	04	25	
2.	45	67	50	29	75	18	72	14	78	95	
3.	62	34	21	48	71	09	73	90	79	62	
4.	89	12	53	36	58	57	86	34	15	24	
5.	98	19	33	62	66	22	21	77	41	05	
6.	53	84	07	00	70	85	19	69	48	73	
7.	700	374	963	864	710	075	112	413	449	042	
8.	591	686	379	516	204	113	044	850	438	567	
9.	430	828	367	855	735	591	338	821	515	376	
10.	439	368	924	474	200	649	912	531	217	052	
11.	425	118	355	984	200	970	572	727	236	726	
12.	979	560	888	085	749	433	354	179	168	989	
13.	4276	2184	8761	9841	6514	6120	1733	8199	8011		
14.	6149	5267	2232	3214	2496	4397	4660	0856	6468		
15.	9164	8811	3225	5170	8653	4717	0161	5124	8245		
16.	1798	0271	7083	5164	4816	3552	4483	0779	9497		
17.	1808	0522	3946	0926	8590	7413	3317	4501	9170		
18.	3655	1041	0454	0009	4961	2374	9625	8546	3127		
19.	69185	96883	77205	44584	18693	03787	11669	45430	25124	05421	
20.	84498	52476	75351	46487	92097	88713	64243	77888	19426	07026	

Figure 9.1. Random digits

If the population is very large, it is often more convenient to draw a random sample by using a table of *random digits* (some refer to it as *random numbers).* To generate a table of random digits, write the numerals 0, 1, 2, 3, 4, 5, 6, 7, 8 and 9, each on a slip of paper and fold the slips so that the digit cannot be seen without unfolding. Mix the slips thoroughly in a hat or a jar and draw one slip. The numeral on that slip is your first random digit. Fold it and put it back. Then draw a second numeral and repeat the process like an independent trial to obtain the desired number of random digits. Each digit you draw should be written on a table in the order it is drawn. For convenience, the table of random digits may be grouped in two's, three's, etc. If you are working with a population of less than 100, the digits may be grouped in pairs and if you are working with a population of less than 1000, the digits may be grouped in three's and so on. The random digit table in Figure 9.1 shows various groupings of the digits. The grouping is to make the table easy to read. We have numbered the rows to make it convenient to refer to. In reality, the table should be regarded as one string of random digits.

In any case, printed tables of random digits are available in texts and it is easy to generate them using a calculator or a computer. Most calculators generate random numbers in three digits. That is fine.

Say we want to draw a sample of 10 from a population of 50 using random digits. We use the following steps:

Step 1.   Label each member of the population with two digits from 00 to 49.

Step 2.   Generate random digits and for each pair of digits select the member with those digits as label. Repeat the process ten times. Discard a digit of 5 or higher if it comes as the first digit and discard any pair of digits that has already been selected.

(Or) If you are using a table, select the members corresponding to successive pairs in the table. Again, discard pairs that give a number higher than 49 or a number that has already been selected.

To select a sequence of 10 pairs, we can start anywhere on the table as long as we take successive digits. Also, it is wise not to use the same starting point all the time. Below are three different strings of 10 random numbers (less than 50) drawn from the table in Figure 9.1, starting at different points.

I.   (starting at the beginning of row 1):   01, 39, 06, 44, 47, 12, 11, 04, 25, 45

II.   (starting at the beginning of row 7):   03, 38, 00, 11, 24, 13, 44, 42, 16, 37

III. (staring at the beginning of row 20):   49, 24, 35, 14, 09, 13, 37, 19, 42, 26

Exercises 9.1

1.   A math teacher wanted to sample the opinion of his class about his teaching. As a sample, he took the 5 top students in the class. Was it a random sample? If your answer is 'no', then give a method for drawing a random sample.

2. A person was asked to sample the average salary of professors at this school. He went to the faculty lounge at noon and selected a random sample of 10 professors who came to have lunch there. Was it a random sample of the professors at this school? If your answer is 'no', state how you can draw a random sample for this survey.

3. The unit of a 'rod' is described as the length of the feet of sixteen men selected at random. In 1514, it was accepted as the length of the feet of the first sixteen men, tall and short, that came out of a church after the service was over. Was it a random sample? If your answer is 'no', give a method for selecting a random sample to measure the length of a 'rod'.

4. State how you would select five names at random from the following list:

Tammy	Peter	John	George	Derek
Sofia	Alvin	Tulip	Paul	Jay
Greg	David	Sandy	Aaron	Tina
Philip	Joy	Gavin	Danny	Sue

5. State how you would pick a random sample of seven two-digit numbers.

6. State how you would pick a random sample of seven families from your town.

7. For the draft, the Selection Service used to call up youths according to birth dates drawn by lottery. Did the youths drafted in a particular year represent a random sample of all young men eligible for the draft that year?

8. To predict the winner in the 1936 presidential election between Roosevelt and Landon, the literary digest magazine sent several million postcards to a sample of US voters drawn from telephone users and car owners. People that answered their cards were heavily in favor of Landon, and the digest predicted a win for Landon. However, Roosevelt won the election a few weeks later. What was wrong with the sample?

9. Take a random sample of 10 students from your class and ask them to write the score they received in the last test of this course. Determine the sample mean and the mean deviation. Compare these with the population (the whole class) mean and the mean deviation for the population.

10. Select a random sample of 15 students from your class using random digits.

11.   A town has 235 households. As part of a national poll, you are required to select at random 4 of these households. State how you would select the 4 households.

12.   A town has 895 households. As part of a national poll, you are required to select at random 5 of these households. State how you would select the 5 households.

2. PUBLIC OPINION POLLS

Public opinion polls have become very much a part of democratic society and an important instrument for the decision-making processes of governments, industries, businesses and public institutions of all kinds. National polls give us an insight to anticipate probable consequences of one decision or another. Now polls are used to gauge public opinion on all kinds of social, political and economic issues and above all they are used to predict election results. Therefore, it is becoming more and more important for a responsible citizen to understand how polls are taken, how they are interpreted, and, if not to challenge them, at least to understand their implications. This is more so because polls are not always correct and various interest groups may manipulate polls or take polls in a manner that would serve their own purpose. Since the methods, the procedures and the principles of opinion polls are based on the mathematical laws of probability, it is possible for a student of finite math to make a critical analysis of opinion polls. Therefore, we discuss some of the basic aspects of polls. For a more detailed account, the reader is referred to the refreshingly enlightening book, "The Sophisticated Poll Watcher's Guide" by George H. Gallup, the founder of Gallup Poll and the American Institute of Public Opinion, published by Princeton Opinion Press (1972). The material here is based mostly on Gallup's book.

Sampling for a National Poll

The most important aspect of polling is the design and selection of the sample. The sample must be random, its size must be large enough to give a broad based cross-section of the national adult population, and it must be drawn at a pertinent time. The method used by the Gallup Poll indicates how these are achieved. The Gallup Poll usually selects a random sample of 1,500 adults (over 18) in the following manner.

Objective 1. Selecting, at random, approximately 300 sampling points.

Organize all the 200,000 or so election districts or precincts in the nation with their population. Find an interval depending upon the number of election districts to be selected. The interval would be the adult population divided by the number of election districts to be selected. If 300 election districts are to be selected, then assuming the adult population to be 210,000,000, the interval would be 700,000.

Starting anywhere at random select the election districts wherever the 700,000 interval falls. These districts are then the 300 sampling points.

Objective 2. Selecting the dwelling unit of the person to be interviewed.

By consulting street maps and block statistics of each district selected, an assistant in the Princeton headquarters of the organization chooses a starting point within the election district at random and instructs an interviewer to select a specified number of people to interview at random.

The interviewer has no control over where to begin his work. The interviewer, following instructions, counts off the number of dwelling units (it may be every third or every fifth or every twelfth dwelling unit) and proceeds to select the person to be interviewed for the opinions.

Objective 3. The interview.

After arriving at the selected dwelling, the interviewer selects an individual at random. A dwelling or a home is the preferred place for interview because a home is where almost all Americans are likely to be found. So, contacting people at home is the basis for all national surveys.

Objective 4. Examining the sample.

As the completed forms are returned from the field to the Princeton office, information about the person interviewed like occupation, age, education, religious preference, etc., are fed into a computer and are examined to see whether the sample selected is a good cross section of the population. This is done by comparing the data provided by the Census Bureau Current Population Surveys on each one of the important factors.

According to Gallup the important factors are

"---the educational level of those interviewed;

---the age level;

---the income level;

---the proportion of males to females;

---the distribution by occupations;

---the proportion of whites to non-whites;

---the geographical distribution cases;

---the city-size distribution.

Typically, when the educational level is correct, that is, when the sample has included the right proportion of those who have attended college, high school, grade school, or no school, when the geographical distribution is right and all areas of the nation have been covered in the correct proportion, when the right proportion of those in each income level has been reached, and the right percentage of whites and non-whites of men and women has been reached - then other factors usually tend to fall in line. These include such factors as religious preference, political party preference, and most other factors that bear upon voting behavior, buying behavior, tastes, interests, and the like."

It is true that other pollsters design their samples differently. The Gallup design gives us a standard to judge the merits of other designs of samples used in a national poll. This ability to judge the merit of the poll samples is worthwhile because pollsters are bound to give this information. In fact, the pollsters are bound by certain principles of disclosure.

Principles of Disclosure

In 1979, the National Council of Public Polls adopted certain Principles of Disclosure by which the pollsters agreed to include reference to the following in all reports of their survey findings:

Sponsorship of the survey;

Dates of interviewing;

Method of obtaining the interviews (in-person, telephone or mail)

Population that was sampled;

Size of the sample;

Size and description of the sub-sample, if the survey report relies primarily on less than the total sample;

Complete wording of questions upon which the release is based; and,

The percentages upon which conclusions are based.

[*Polling On The Issues,* edited by Albert H. Cantril and published by Seven Locks Press, 1980].

Poll by Phone

Many opinion polls are conducted by telephone. After individuals are selected at random, the selected individual's interview is taken by telephone. For example, below we give the details of a New York Times/CBS News poll conducted in March 6, 1991.

"The sample of telephone exchanges called was selected by a computer from a complete list of exchanges in the country excluding Alaska and Hawaii. The exchanges were chosen to insure that each region of the country was represented in proportion to its population. For each exchange, the telephone numbers were formed by random digits, thus permitting access to listed and unlisted numbers. The numbers were then screened to limit calls to homeowners.

The results have been weighted to take account of household size and number of telephone lines into the home and to adjust for variations in the sample relating to region, race, sex, age and education. The sample size was 1,252 and the confidence level was 95 percent."

Now that almost 95% of American households have a phone, poll by phone has some advantages. It is comparatively easier, more convenient, and faster than the laborious method described by Gallup. Poll by phone also has the advantage over written answers because it is easier to detect if a person is lying.

One problem in conducting poll by phones is that 30% of American residential phones are unlisted. Therefore, one cannot use a telephone directory to select a random sample of telephone numbers. Gallup follows the following prcedure to select a random sample of households for a poll by phone.

(1) Collect a computerized list of all telephone exchanges in the country along with estimates residential households attached to each exchange.

(2) Prepare telephone samples from that list with the help of the computer. This is done by using a procedure called the Random Digit Dialing (RDD). Thus it creates a list of household telephone numbers in America.

(3) Select a random sample from the list in (2) for the interviewers to call.

(4) Once the household is reached by phone, Gallup tries to select at random an adult in the house and start the interview. This may be done by obtaining from the person who answers the phone the names of the adults in the house along with their gender and age. Gallup then selects at random one of those persons for the interview and requests him or her to come to the phone.

(5) What if the person selected for the interview is not at home or the number called is busy or there is no answer? Usually, the number is called back again and again until contact is made within the period of the survey. Of course, all this is done by using the computer where the number not reached is stored for redialing.

Size of Sample

Should the size of a sample depend on the size of the population? In other words, if we are surveying opinion poll in New York should, should we select a sample 10 times smaller (since the population of the New York state is tenth of the population of the country) than the 1,500 used for a national poll? The important character of the random sample is that its size is designed to get a good cross-section of the people based on the important factors. Therefore, the size of the sample should be the same whether it is for a state poll or a national poll. Gallup gives the following example to explain this mystifying fact: "Suppose that a hotel cook has two kinds of soup on the stove - one in a very large pot, another in a small pot. After thoroughly stirring the soup in both pots, the cook need not take a greater number of spoonful from the large pot or fewer spoonful from the small pot to taste the quality of the soup, since the quality should be the same."

The Timing

The timing of the poll is underscored by the following historic cases.

1. To predict the winner in the 1948 presidential election between Dewey and Truman, Gallup took a sample opinion early in October and predicted Dewey would win. In the November election, Truman won. What was wrong with the sample?

   This sample was drawn too early and failed to reflect the definite shift by the electorate to Truman during the closing days of the election. Pollsters do not make that kind of mistake nowadays.

2. In the 1976 election, a poll in the summer showed Carter ahead of Ford by 33 percentage points, one of the largest leads in polling history during a presidential campaign. However, Ford was able to cut the lead and the final 2-point victory of Carter was not assured until all the votes were counted. Therefore, a prediction based on a poll drawn too early could not have been accurate.

Figure 9.2 gives a table of Gallup poll predictions and the actual votes received by each winning candidate in the more recent presidential elections. The accuracies reflect what can be achieved by proper sampling. According to Gallup organization, "Gallup's final estimate in each presidential election is based on likely voters, and takes into account assumptions about the vote preference of undecided voters.

Year	Final Gallup Winner	Prediction (%) predicted	popular votes received	Error
2000	Bush	48	48	0.0
1996	Clinton	52	49.2	2.8
1992	Clinton	49	43	6.0
1988	Bush	56	53.4	2.6
1984	Reagan	59	58.8	.2
1980	Reagan	47.0	50.8	-3.8
1976	Carter	48	50.1	-2.1
1972	Nixon	62	60.7	1.3
1968	Nixon	43	43.4	-.4
1964	Johnson	64	61.3	2.7
1960	Kennedy	51	49.9	1.1
1956	Eisenhower	59.5	57.6	1.9
1952	Eisenhower	51	55.1	-4.1
1948	Truman	44.5	49.6	-5.1
1944	Roosevelt	51.5	53.6	-2.1
1940	Roosevelt	52	54.7	-2.7
1936	Roosevelt	56	60.8	-4.8

Figure 9.2. Record of Gallup Poll Accuracy in Presidential Elections

Source: *The Gallup Organization, Princeton. Website www.gallup.com*

Margin of Error

Opinion polls make predictions for the future. Like any other probability estimates, it is reported with a confidence level and a margin of error. The confidence level is usually chosen to be 95 percent and the margin of error is usually not more than ± 3 percentage points. These are figured by using various statistical methods. Gallup gives the following table for guidelines:

Percentages	Sample size						
	1500	1000	750	400	300	200	100
near 10	2	2	3	3	4	5	7
near 20	2	3	4	4	5	7	9
near 30	3	4	4	4	6	8	10
near 40	3	4	4	5	6	8	11
near 50	3	4	4	5	6	8	11
near 60	3	4	4	5	6	8	11
near 70	3	4	4	4	6	8	10
near 80	2	3	4	4	5	7	9
near 90	2	2	3	3	4	5	7

Figure 9.3. Table to determine margin of error

Suppose a reported percentage is 37 for a sample of size 1,000. Since 37 is near 40 in the table, look along the percentages near 40 and under sample size 1,000. You'll see that the margin of error is 4. This means the poll is 95 percent confident that the true percentage is between 37 - 4 = 33 and 37 + 4 = 41.

The Questions

The wording of the questions, the order of the questions are an important part of a poll. If they are not carefully selected they may be a source of bias and error in the data. For example, in presidential race, should the name of the vice presidential candidates be inluded along with the presidential candidate or not? Should the party affiliation of the candidates be mentioned or not? Gallup poll does include the names of the vice presidential candidates and the party affiliations since the voter would see these information when reading the ballot in the voting booth. "Gallup's rule in this situation is to ask the question in a way which mimics the voting experience as much as possible".

Questions about policy issues and opinions about social or educational issues like gun control, abortion, prayers in schools demand even more care. This is where the track record, the integrity and the creativity of the pollster come to play an important role. For more about this, please see Frank Newport, Lydia Saad, David Moore, *Where America Stands*, 1997 John Wiley / Sons.

Exercises 9.2

1. Senator Albert Gore, of Tennessee once commented, "As a layman I would question that a straw poll of less than 1 percent of the people could under any reasonable circumstance be regarded as a fair and meaningful cross section. This would be something more than 500 times as large a sample as Dr. Gallup takes." State whether Senator Gore's criticism is justified.

2. Compare the two methods of sampling, namely the method of personal interviews as detailed by Gallup and sampling by telephone calls as detailed by New York Times/CBS News Poll.
   What could be the advantages and disadvantages of polls made by telephone?

3. What is the probability that an American adult will be selected in a random sample of 1,500 [Assume the adult population of America to be 210,000,000].

4. The Washington correspondent of a large Midwest daily once commented, "In more than 40 years of newspapering around this country, and the world for that matter, I have never met anyone who admitted being polled by a public opinion poll and I have met a lot of people." Criticize this comment in light of the discussions presented here.

5. Most members of Congress conduct their personal polls by mail. They send out cards with a set of questions and decide everything from the answers that they receive back. Discuss the merits of such a poll by mail.

6. A large cask contains 70,000 white and 30, 000 black balls and a smaller cask contains 700 white and 300 black balls. After mixing the balls thoroughly in either cask, a blindfolded person, is asked to draw exactly 100 balls out of each cask. Is it likely that the proportion of the white balls to the black balls will be different in the two samples drawn by the person?

7. In a poll during the Persian Gulf war it was found that 64% of adult Americans feel it is unacceptable for the United States to send women with young children to the war zone. If the survey had a confidence level of 95% and was based on a random sample of 1,000, find the margin of error using the table in Figure 9.3.

8. In a poll during the Persian Gulf war it was found that 28% of adult Americans feel it is unacceptable for the United States to send men with young children to the

war zone. If this survey had a confidence level of 95% and was based on a random sample of 1,000, find the margin of error using the table in Figure 9.3.

9.   According to the New York Times/CBS News Poll conducted on March 6, 1991, 83% of the American people approved of the way President Bush was handling foreign policy. This poll had a confidence level of 95 percent and a sample size of 1,252. Determine the margin of error by using the method given in section 5 of chapter 7. Round the errors to the closest percentage point and give the 95 percent confidence interval.

10.   According to the New York Times/CBS News Poll conducted on March 6, 1991, 42% of the American people approved of the way President Bush was handling economic policy. This poll had a confidence level of 95 percent and a sample size of 1,252. Give the 95 percent confidence interval for the true percentage of American people who supported President Bush's handling of the economy.

11.   Are college students living on campus, armed forces personnel living on military bases, prisoners or hospital patients reached in the public opinion poll conducted by Gallup?

12.   Give a reason why telephone books cannot be used to select a sample of households in an area?

3. SAMPLING DISTRIBUTION

Assume that there are 20,000 students at a college out of which 11,900 are females. Then the population proportion p of females is $\frac{11900}{20000} = \frac{119}{200}$. If a random sample of 40 students contain 28 females, then the sample proportion $\bar{p}$ of females is $\frac{28}{40} = \frac{7}{10} = .7.$ Table in Figure 9.4 gives the sample proportion of 10 samples each of size 40.

If we draw all possible random samples of size 40 from the students, then each possible value of $\bar{p}$ can be assigned a probability as shown in Figure 9.5. Therefore, the sample proportion $\bar{p}$ is a random variable. The probability distribution of $\bar{p}$ is called the *sampling distribution of* $\bar{p}$. This distribution has the following properties:

1.   The mean of $\bar{p}$ = p.

2. By the Central Limit Theorem, the sampling distribution of $\bar{p}$ is approximately normal for sufficiently large sample size. (The sample size is considered large if both np and n(1 - p) are greater than 5, that is, np > 5 and n(1 - p) > 5.)

3. The standard deviation $\sigma_{\bar{p}}$ of the sample proportion is given by:

$$\sigma_{\bar{p}} = \sqrt{\frac{p(1-p)}{n}}, \tag{1}$$

where p is the population proportion and n is the sample size.

Sample No.	Sample Proportion
1.	.70
2.	.66
3.	.63
4.	.63
5.	.69
6.	.66
7.	.70
8.	.64
9.	.65
10.	.64

Figure 9.4

Mean	Frequency	Probability
.63	2	2/10
.64	2	2/10
.65	1	1/10
.66	2	2/10
.69	1	1/10
.70	2	2/10

Figure 9.5

Example 9.2

1. Forty percent of the students at a university are females. Let $\bar{p}$ be the proportion of females in a random sample of 36 students. Find the mean and the standard deviation of $\bar{p}$.

Solution. p = 40% = .4, 1 - p = .6. So, the mean of $\bar{p}$ = .4 and standard deviation of

$$\bar{p} \text{ is } \sigma_{\bar{p}} = \sqrt{\frac{p(1-p)}{n}} = \sqrt{\frac{.4 \times .6}{36}} = .0816.$$

Exercises 9.3

1. Twenty percent of the students at a university are females. Let $\bar{p}$ be the proportion of females in a random sample of 36 students. Find the mean and the standard deviation of $\bar{p}$.

2. Fifty percent of the students at a university are females. Let $\bar{p}$ be the proportion of females in a random sample of 49 students. Find the mean and the standard deviation of $\bar{p}$.

3. Sixty percent of the voters in a state are democrats. Let $\bar{p}$ be the proportion of democrats in a random sample of 96 voters of that state. Find the mean and the standard deviation of $\bar{p}$.

4. Thirty percent of the voters in a state are democrats. Let $\bar{p}$ be the proportion of democrats in a random sample of 84 voters of that state. Find the mean and the standard deviation of $\bar{p}$.

5. Twenty percent of the students on a campus do not like math. Let $\bar{p}$ be the proportion of students who do not like math in a random sample of 100 students of that campus. Find the mean and the standard deviation of $\bar{p}$.

Estimating population proportion from sample proportion

The population proportion is a population parameter. In general, a number that is calculated for a population is called a *parameter*. A number that is calculated for a sample is called a *statistic*. Thus $\bar{p}$ is an example of a statistic. In case of very large population, like the voters of a country, we need to estimate a parameter from one or more random samples using the sample statistic. If only one sample is used then the value of the sample statistic used is called a *point estimator* of the population parameter. The point estimate is bound to contain some error. However, since the distribution is normal or approximately normal, we can calculate the margin of error using properties of normal distribution. The margin of error will depend upon how confident we wish to be in our estimate – or in our confidence level. The *confidence level* is expressed in the form of a percentage. For example, 95% confidence level means the probability that the true value is contained in that interval is .95. This also means in 95 cases out of 100, our estimate will prevail and in the remaining 5 the estimate may fail.

The calculation of the margin of error requires how many standard deviation we should allow from the point estimate. So, it will be in the form of $z \; \sigma_{\bar{p}}$, where $\sigma_{\bar{p}}$ is the standard deviation of the estimate and z is the z-value of the deviation to be allowed. The formula for the margin of error is

$$\text{Margin of error} = z \; \sigma_{\overline{p}} = z \times \sqrt{\frac{\overline{p}(1 - \overline{p})}{n}}, \tag{2}$$

where $\overline{p}$ is the sample proportion, n is the sample size, and z is the z-value for the confidence level.

Since it is just a question of estimate, we may replace the term $\sqrt{\overline{p}(1 - \overline{p})}$ by its largest possible value $\frac{1}{2}$ [see note at the end of this chapter]. This results in the following simpler formula for the Margin of error :

$$\text{Margin of error} = \frac{z}{2\sqrt{n}}. \tag{3}$$

The estimate for the population parameter p is then given in the form of an interval within which we believe the true p lies. The interval is given by:

$$[\overline{p} - \frac{z}{2\sqrt{n}}, \quad \overline{p} + \frac{z}{2\sqrt{n}}] \tag{4}$$

[Sample proportion - Margin of error,   Sample proportion + Margin of error].

The interval given in (4) is the *Confidence Interval* for p. This depends on the confidence level we are seeking.

Examples 9.3

1. A random sample of 36 students at a university had 21 females. Find
    (a) The point estimate of the proportion of females on the campus.
    (b) 95% confidence interval for proportion of females on the campus.

Solution.   (a) $\frac{21}{36} = \frac{7}{12} = .58$ or 58%

    (b) To answer this part we proceed as follows:

Step-1. Determine the z-score for the confidence level of 95%. For 95% confidence level, we will have to find z such that the area under the normal curve between -z and +z is .95. That means the area between 0 and z is .95/2 = 0.475. From the table of areas under the normal curve, the value of z = 2 (approximately).

Step-2. Calculate the Margin of error using formula (3): Margin of Error = $\frac{z}{2\sqrt{n}}$.

In this case, $z = 2$ as determined in step-1, $n = 36$. So,

$$\text{Margin of error} = \frac{2}{2 \times 6} = \frac{1}{6} = .17 .$$

Step-3.   So, the confidence interval is $[.58 - .17, \ .58 + .17]$ or $[.41, .75]$.

Gallup's table to determine margin of error (Figure 9.3)   for opinion poll is perhaps based on formula (2), namely,

$$z \ \sigma_{\bar{p}} = z \ \sqrt{\frac{\bar{p}(1 - \bar{p})}{n}}$$

where $\bar{p}$ is the sample proportion. This formula depends on $\bar{p}$ as well as the sample size n. In the table $\bar{p}$ is given in terms of percentages. Thus to find the margin of error when percentage is near 20 and $n = 1500$, we take $\bar{p} = .2$ and $z = 2$, (see 1 of examples 9.3) for 95% confidence interval. Then

$$2 \times \sqrt{\frac{.2 \times .8}{1500}} = .020, \text{ which in percentage point equals 2.}$$

Similarly, near 60 percentage and $n = 1000$, we get

$$2 \times \sqrt{\frac{.4 \times .6}{1000}} = .031 \text{ which is estimated upwards to 4 percentage points.}$$

Similarly, near 60 percentage and $n = 750$, we get

$$2 \times \sqrt{\frac{.4 \times .6}{750}} = .036 \text{ which is estimated upwards to 4 percentage points.}$$

Similarly, near 60 percentage and $n = 400$, we get

$$2 \times \sqrt{\frac{.4 \times .6}{400}} = .048 \text{ which is estimated upwards to 5 percentage points.}$$

Note that increasing sample size reduces the margin of error.

Exercises 9.4

1.   Find the margin of error $\dfrac{z}{2\sqrt{n}}$ for 95 percent confidence interval when the sample size is

(a) 100   (b) 900   (c) 1200   (d) 1500   (e) 1600

2. Find margin of error $\dfrac{z}{2\sqrt{n}}$ for 90 percent confidence interval when the sample size is (a) 100   (b) 800   (c) 1200   (d) 1500   (e) 1600

3. A random sample of 49 students at a university had 21 females. Find
   (a) The point estimate of the proportion of females on the campus.
   (b) 95% confidence interval for proportion of females on the campus.

4. A random sample of 100 students at a university had 40 females. Find
   (a) The point estimate of the proportion of females on the campus.
   (b) 99% confidence interval for proportion of females on the campus.

5. (a) Does the margin of error depend on the size of the population or on the size of the sample?
   (b) If the population gets larger, will the margin of error get larger or smaller?
   (c) If the sample size gets larger, will the margin of error get larger or smaller?
   (d) If a larger confidence level is sought, will the margin of error get larger or smaller?

6. Find the proportion of females in your class. Assuming that your class is a random sample of students on your campus, estimate the 90% confidence interval for proportion of female students on your campus.

7. A random sample of 64 voters of a state contained 48 democrats. Find
   (a) The point estimate of the proportion of democrats in that state.
   (b) 80% confidence interval for the proportion of democrats in that state.

8. A random sample of 100 voters of a state contained 70 democrats. Find
   (a) The point estimate of the proportion of democrats in that state.
   (b) 90% confidence interval for the proportion of democrats in that state.

9. Verify the entries in Gallup's table when $\bar{p}$ is near 50 percent and $n = 1500$, $n = 1000$, $n = 750$ and $n = 400$.

10. Make a table to determine the margin of error using formula (2) when the confidence level is 99% and for the following values of $\bar{p}$ and n:

    $\bar{p}$ = near 20%, near 40% and near 50%;

    $n = 1500, 1000, 750, 400, 300, 200,$ and 100 .

4. PROBABILITY ESTIMATES

If an independent trial with two outcomes and with probability p for success is repeated n times, then the expected number of successes is equal to np. So,

$$p = \frac{\text{Expected number of successes}}{\text{Number of times the trial is repeated}}.$$

The actual number of successes usually deviates from the expected number np. If x is the acual number of successes, the number x/n will not necessarily be the value of p. In fact, one can never be 100 percent sure that an estimated probability is the right probability. However, by using the central limit theorem, one can determine an interval containing the number x/n depending upon how sure one wishes to be of being correct. In this case also, we have to consider a confidence interval as illustrated in Examples 9.3.

Example 9.4

Suppose an independent trial with two outcomes is repeated 10,000 times and success is observed 6,000 times. What is the 95 percent confidence interval for the probability of success on each trial?

Solution. The solution to this problem is [0.59, .61], which means that with 95 percent certainty one can say that the required probability is between 0.59 and .61. The 95 percent confidence implies that if the trial is repeated, then in 95 cases out of 100 the observed probability will be within the estimated probability [0.59, .61]. The steps below show how we arrived at this solution.

Step 1.   For a 95 percent confidence interval, $A(z) = 0.475$, which gives $z = 2$ (approx.) [see 1 of examples 9.3].

Step 2.   Find the *margin of error* by using the following formula.

$$\text{Margin of error} = \frac{z}{2\sqrt{n}}$$

In the present problem, $\text{Margin of error} = \frac{2}{2\sqrt{10000}} = 0.01.$

Step 3.   Find $\dfrac{x}{n} = \dfrac{\text{Actual number of successes}}{\text{Number of times the trial is repeated}}.$

In this case, $x/n = 6{,}000/10{,}000 = 0.6$

Step 4.   The 95 percent confidence interval for the probability of success is then given by

$$[0.6 - \frac{z}{2\sqrt{n}}, \quad 0.6 + \frac{z}{2\sqrt{n}}] = [0.6 - 0.01, \quad 0.6 + 0.01]$$

In general,

The b percent confidence interval for the probability of success of an independent trial with two outcomes is given by

$$[\frac{x}{n} - \frac{z}{2\sqrt{n}}, \quad \frac{x}{n} + \frac{z}{2\sqrt{n}}],$$

where x = actual number of successes, n = number of times the trial is repeated, z is the z-value determined from the table for areas under the normal curve and the formula

$$A(z) = \frac{1}{2} \times \frac{b}{100}.$$

The end points of a confidence interval are the *confidence limits* of the *confidence level* (95 percent in this problem). The following theorem gives the justification for the definition of confidence interval and for the method used in solving the above problem.

THEOREM 9.1. *Let an independent trial with two outcomes and with probability p for success be repeated n times. If the observed number of successes x satisfy*

$$-z \leq \frac{x - np}{\sqrt{np(1 - p)}} \leq z, \quad then \quad p \in [\frac{x}{n} - \frac{z}{2\sqrt{n}}, \quad \frac{x}{n} + \frac{z}{2\sqrt{n}}].$$

Proof.   If $-z \leq \dfrac{x - np}{\sqrt{np(1 - p)}} \leq z$, then multiplying each term by $\sqrt{np(1 - p)}$, we get

$$-z\sqrt{np(1 - p)} \leq x - np \leq z\sqrt{np(1 - p)}$$

Dividing each term by n,

$$-z \frac{\sqrt{p(1 - p)}}{\sqrt{n}} \leq \frac{x}{n} - p \leq z \frac{\sqrt{p(1 - p)}}{\sqrt{n}}.$$

We know that for any number p [see note at the end of this chapter],

$$\sqrt{p(1 - p)} \leq \frac{1}{2}.$$

Using this fact, we get

$$-z\left(\frac{1}{2\sqrt{n}}\right) \leq \frac{X}{n} - p \leq z\left(\frac{1}{2\sqrt{n}}\right)$$

This implies

$$-\frac{z}{2\sqrt{n}} \leq p - \frac{X}{n} \leq \frac{z}{2\sqrt{n}}, \quad \text{that is,} \quad \frac{X}{n} - \frac{z}{2\sqrt{n}} \leq p \leq \frac{X}{n} + \frac{z}{2\sqrt{n}}.$$

From which we can conclude that $p \in [\frac{X}{n} - \frac{z}{2\sqrt{n}}, \ \frac{X}{n} + \frac{z}{2\sqrt{n}}]$. ∎

Corollary. For any given z, the probability that p will lie in the interval

$$[\frac{X}{n} - \frac{z}{2\sqrt{n}}, \ \frac{X}{n} + \frac{z}{2\sqrt{n}}]$$

is approximately equal to the area under the normal curve between -z and z.

We have:

$$\Pr\left[p \in [\frac{X}{n} - \frac{z}{2\sqrt{n}}, \ \frac{X}{n} + \frac{z}{2\sqrt{n}}]\right] \qquad \Pr[-z \leq \frac{X - np}{\sqrt{np(1-p)}} \leq z]$$

Hence, the corollary follows from the central limit theorem.

The quantity $\frac{z}{2\sqrt{n}}$ as given in Theorem 9.1 is the *the margin of error.* Here n represents the number of trials. If the number of trials is larger than the margin of error would be smaller.

Exercises 9.5

1.   A defective coin is tossed 400 times and it comes up heads 100 times. What is the 95 percent confidence interval for the probability of getting a head each time the coin is tossed?

2.   A visitor at the University of Rhode Island wanted to estimate the proportion of female students at the University. He found out that there were 40 girls out of a total of 100 students at the union. What would be the 90 percent confidence interval for the proportion of female students at the University of Rhode Island?

3. A die is rolled 100 times. It comes up with sixes 20 times. What is the 99 percent confidence interval for the probability of getting a six each time the die is rolled?

4. An opinion poll on a sample of 900 voters showed that 40 percent are in favor of candidate A. Find the 99 percent confidence limits on the true percentage of the population in favor of candidate A.

5. A man needing dental surgery decides that he would undergo the surgery if he could be 95 percent sure that there was at least a 0.65 probability of success. The dental surgeon tells him that the surgery has been successful 50 out of 64 times. Does the man agree to have the surgery?

6. If the confidence is increased (say from 95 percent to 99 percent), does the length of the confidence interval get smaller or larger? Verify this by finding the 99 percent confidence interval in problem 1.

7. Find the 80 percent confidence interval in problem 3. If the confidence is decreased, does the length of the confidence interval get larger or smaller?

5. ESTIMATING POPULATION SIZE

Example 9.5. You are asked to estimate the number of trout in a lake. Give a method to determine a 95 percent confidence interval for the total number of trout in the lake.

Solution. This may be approached very much like an independent trial with two outcomes. Let N be the number of trout in the lake. It is required to obtain an estimate for N which has a 95 percent chance of being correct.

Step 1. Take out a sample of, say, 100 trout at random and put them back after tagging. Then,

$$\frac{\text{Number of tagged trout}}{\text{Total number of trout}} = \frac{100}{N}$$

Step 2. After giving sufficient time for the tagged trout to mix freely with the rest, take out another random sample of, say, 900 trout. Count the number of tagged trout in the sample, say it is 70. Then,

$$\frac{\text{Number of tagged trout in sample}}{\text{Number in sample}} = \frac{70}{900} = \frac{7}{90}.$$

Step 3.   Find z as in step 1 of Example 9.2, so that z = 2. Calculate the margin of error by the formula

$$\frac{z}{2\sqrt{n}} = \frac{z}{2\sqrt{\text{Number in sample}}} = \frac{2}{2\sqrt{900}} = \frac{1}{30}.$$

Step 4.   The 95 percent confidence interval for N is then given by

$$\frac{7}{90} - \frac{1}{30} \leq \frac{100}{N} \leq \frac{7}{90} + \frac{1}{30}$$

$$\frac{4}{90} \leq \frac{100}{N} \leq \frac{10}{90} \text{ , that is, } 4 \leq \frac{9000}{N} \leq 10$$

Step 5.   Taking the reciprocal of every term of the inequlity in step 4, we get

$$\frac{1}{10} \leq \frac{N}{9000} \leq \frac{1}{4}$$

Multiplying every term by 9,000, we get $900 \leq N \leq 2250$.

Therefore, with 95 percent surety, one can say that the number of trout in the lake is between 900 and 2,250.

Note 1. In step 5 we have used the rule that if every term in an inequality is changed to its reciprocal, then the inequality reverses direction. For example,

$$\frac{1}{10} < \frac{1}{4} \text{ because } 10 > 4.$$

Note 2. In the margin of error $\dfrac{z}{2\sqrt{n}}$, n represents the sample size. So, the margin of error does not depend on the population size, but on the sample size.

Exercises 9.6

1.   100 fish from a pond are tagged. A later sample of 400 fish from the pond contains 30 tagged fish. Find the 95 percent confidence interval for the total number of fish in the pond.

2.   A wholesale apple dealer mixes rotten apples with good ones, then puts them at random in boxes of hundreds and ships them out. An official discovered that in

one batch he had mixed one thousand rotten apples. The official was able to get a box of one hundred apples which contained 20 rotten apples. Give a 95 percent confidence interval for the total number of apples the dealer had.

3.   Give the 99 percent confidence interval for the number of fish in problem 1.

4.   Give a method to estimate the number of deer in a forest.

6. TEST OF HYPOTHESIS

Statistics gives methods to choose between two or more competing proportions. While comparing probabilities, it is convenient to denote the two probabilities by p_0 and p_1. We illustrate the general method with Examples 9.4.

Examples 9.6

1.   A math professor grades his students on the basis of his own assessment of how much a student learns in the course. Once, a student who received a 'C' (70%) challenged the professor because the student claimed to know 80% (B) of the material. How can you decide, with 90 percent surety, whether to accept the student's claim or the professor's claim?

Solution.   Here we have two competing probabilities about the student's ability to answer a question, namely,

$$p_0 = 0.7 \text{ according to the professor}$$
$$p_1 = 0.8 \text{ according to the student}$$

The idea is to devise a test and to settle the issue one way or the other. Therefore, we must decide how many questions should be in the test and how many of these the student should be able to answer correctly in order to establish his claim. Let n be the number of questions on the test, and let x be the number of questions that the student must answer correctly. We have to determine n and x.

No matter how careful we are, there is always an element of some error in the sense that we may agree with the student even though the professor is correct or we may agree with the professor even though the student is correct. We can reduce the probability of the error to an acceptable level (in this case, .10).

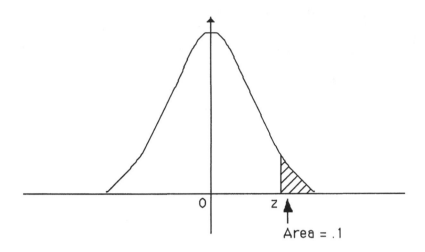

Figure 9.4

From the professor's point of view, he expects np_0 questions to be answered correctly. On the basis of the agreed level of error, the professor should be allowed z units of standard deviations such that the professor's claim will not be accepted if $x > np_0 + z\sqrt{np_0(1 - p_0)}$.

Since we wish this to happen with probability of 0.1, we must take z so that

$$\Pr[x > np_0 + z\sqrt{np_0(1 - p_0)}] = 0.10.$$

Therefore, $0.5 - A(z) = 0.10$, which gives $A(z) = 0.4$ and $z = 1.3$.

From the student's point of view, he expects to answer np_1 questions correctly. On the basis of the accepted level of error, he should be allowed the same amount of standard deviations as the professor, namely $z = 1.3$. Therefore, we will reject his claim if $x < np_1 - z\sqrt{np_1(1 - p_1)}$. This gives

$$np_0 + z\sqrt{np_0(1 - p_0)} < x < np_1 - z\sqrt{np_1(1 - p_1)}$$

From this we can obtain a condition on n in the following manner:

$$np_1 - z\sqrt{np_1(1 - p_1)} > np_0 + z\sqrt{np_0(1 - p_0)}$$

$$\text{or}\ \ n(p_1 - p_0) > z[\sqrt{np_0(1 - p_0)} + \sqrt{np_1(1 - p_1)}]$$

$$\text{Or} \qquad n > \frac{z\sqrt{n}\ [\sqrt{p_0(1 - p_0)} + \sqrt{p_1(1 - p_1)}]}{p_1 - p_0}.$$

Dividing both sides by $\sqrt{n}$, we get

$$\sqrt{n} \; > \; \frac{z[\sqrt{p_0(1 - p_0)} + \sqrt{p_1(1 - p_1)}]}{p_1 - p_0}.$$

Squaring, we get the following general formula for n:

$$n \; > \; \frac{z^2[\sqrt{p_0(1 - p_0)} + \sqrt{p_1(1 - p_1)}]^2}{(p_1 - p_0)^2}$$

In the present problem, $z = 1.3$, $p_0 = 0.7$, and $p_1 = 0.8$. Hence,

$$n \; > \; \frac{(1.3)^2[0.4583 + 0.4]^2}{(0.1)^2}$$

$$> \; 124.499.$$

Therefore, we must take $n = 125$, which means that the test should contain 125 questions. To decide how many questions the student should be able to answer correctly to establish his claim, we determine the value for x as follows:

$$np_0 + z\sqrt{np_0(1 - p_0)} = 125 \times 0.7 + (1.3) \times 26.25 = 94.16.$$

Since $x > np_0 + z\sqrt{np_0(1 - p_0)}$, it follows that the student must be able to answer at least 95 questions correctly. If he answers less than 95 questions correctly, then we will accept the professor's claim because

$$np_1 - \sqrt{np_1(1 - p_1)} = 125 \times 0.8 - (1.3) \times \sqrt{20} = 94.1862$$

The decision made on the basis of this test has a probability 0.90 of being correct.

This method of deciding between two competing proportions is known as testing of hypotheses. Usually the smaller probability (p_0 in example) is referred to as the *null hypothesis* and the larger probability (p_1 in example) is referred to as the *alternate hypothesis*.

2.   It is projected that the average incidence of flu in New England next winter will be 500 per 100,000. A person in this area will, therefore, have a probability 0.005 of catching the flu. The manufacturer of a flu vaccine claims that a person vaccinated

will have only 0.0025 probability of catching the flu. Design a test to verify the claim of the manufacturer with 95 percent certainty.

Solution.   Null hypothesis $p_0 = 0.0025$

Alternate hypothesis $p_1 = 0.005$

Step 1.   Find z from table for accepted margin of error using the formula $A(z) = 0.5$ - margin of error.

$A(z) = 0.5 - (1 - 0.95) = 0.45$, and so, $z = 1.6$.

Step 2.   Calculate the *critical ratio* r given by (see Example 9.4)

$$r = \frac{\sqrt{p_0(1 - p_0)} + \sqrt{p_1(1 - p_1)}}{p_1 - p_0}.$$

In this case, $r = \dfrac{\sqrt{.0025 \times .9975} + \sqrt{.005 \times .995}}{.0025} = 48.19.$

Step 3.   Find n so that $n > z^2 r^2$, where z is from step 1 and r is from step 2.

Hence $n > (1.6)^2(48.19)^2$, that is $n > 5944.6459$.

Step 4   Find $np_0 + z\sqrt{np_0(1 - p_0)}$ or $np_1 - z\sqrt{np_1(1 - p_1)}$. Here the first quantity is 21.0225.

Therefore, a total of 5,945 persons should be vaccinated. If less than 22 of them get the flu, then we accept the claim of the manufacturer. Otherwise, we reject the claim.

Exercises 9.7

1.   A college professor grades his students on the basis of the percentage of material the students learn in his course. Once a student who received a D (60%) challenged the professor claiming that he knew at least 70% of the material and should have received a better grade. Design a test that will settle the dispute with a probability 0.90.

2.   Do problem 1 when the dispute must be settled with a probability of 0.95.

3.   It is projected that the average incidence of flu in New England next winter will be 1,000 per 100,000. The manufacturer of a flu vaccine claims a person vaccinated

will have only 0.0001 probability of catching the flu. Design a test that would decide the claim of the manufacturer with 95 percent certainty.

4. Do problem 3 when the dispute must be settled with a probability of 0.99.

5. Before the Presidential election, an opinion pollster finds that 45 percent of the voters are in favor of candidate A while candidate A claims 50 percent of the voters to be in his favor. Design a test that will settle the argument with probability .99.

CHAPTER TEST 9

1. Find the margin of error for a confidence level of 89% and a sample size of n = 1,600.

2. Outline a method to select a random sample of 1,200 adults from your state.

3. A poll in March, 1991, indicated that only 35% of Americans feel that George Bush has made progress in protecting the environment. If the sample size for this poll was 1,200, find the margin of error using Gallup's table in Figure 9.3.

4. According to Gallup, what factors are important to get a broad based cross-section of the American people?

5. (a) An interviewer for a poll was to select 5 persons at random from a high rise apartment. Since he did not like to ride on elevator, he decided to interview the first five persons that were coming out of the building. Would that adversely effect the sampling process?

    (b) An interviewer for an opinion poll was to select one person at random from a dwelling. The dwelling had three adults. So, he asked whose birthday was coming up and selected that person. Would that adversely effect the sampling process?

6. A table of random digits has 5,000 digits.

    (a) What is the expected number of 1's in the table?

    (b) If you select a random sample of 10 three digit numbers from that table, what is the expected number of 0's in the sample selected?

7.   An opinion poll on a sample of 10,000 voters showed that 80% are in favor of candidate A. Find the 90 percent confidence limits on the true percentage of the population in favor of candidate A.

8.   100 fish from a lake are tagged. A later sample of 1,600 fish from the lake contained 20 tagged fishes. Find the 95 percent confidence interval for the total number of fish in the lake.

9.   A student, who received a 'B' in a course, challenges his professor, claiming that he should get an 'A' since he learnt 90% of the material taught in the course. His professor insists that he has learnt only 80% of the material covered in the course. Design a test that will settle the dispute with 99 percent certainty.

10. The probability of a person catching the flu is 0.005. The manufacturer of a vaccine claims that for a vaccinated person this will reduce to 0.001. Design a test to decide the claim of the manufacturer with 95 percent certainty.

Note. How to show that for any number p, $\sqrt{p(1-p)} \le \frac{1}{2}$.

It is understood that since a negative number do not have a square root, $p(1-p)$ must be positive. So, p and $(1-p)$ must be both positive or both negative. If p and $(1-p)$ are both positive then $0 < p < 1$. This is the case for us.

Proof - 1 (graphically). Tabulate $p(1-p)$ for various positive values of p and less than 1 and graph. You would notice that for $p = .5$ the value of $p(1-p)$ is .25 but for any other value of p it is less than .25.

Proof - 2: (algebraic)   Taking $p = \frac{1}{2} + y$, we get

$$p(1-p) = (\frac{1}{2} + y)(\frac{1}{2} - y) = \frac{1}{4} - y^2.$$

Therefore, $p(1-p) \le \frac{1}{4}$ or $\sqrt{p(1-p)} \le \frac{1}{2}$.

Proof - 2 (applying calculus): Let $f(p) = p(1-p) = p - p^2$. Then $f'(p) = 0$ gives $1 - 2p = 0$, that is, $p = 1/2$ is a critical value of $f(p)$. By taking the second derivative it follows that $p = 1/2$ gives the maximum value of $f(p)$.

Chapter 10

FUN PROBLEMS

There are many historic and interesting problems that one can tackle with a knowledge of the materials in this text. Some of the problems need ingenious proofs but they are all fun. These problems have been around in so many different forms that it is difficult to find the original sources. However, the book by Yaglom and Yaglom and the book by Moesteller are good sources for some of these and other problems.

Exercises 10.1

1.   Of 9 coins of the same denomination, 8 weigh the same and 1 is heavier. Show how you can find the heavier coin in two weighings on a balance scale without using weights.

2.   Of 8 coins of the same denomination, 7 weigh the same and 1 is heavier. Show how you can find the heavier coin in two weighings on a balance scale without using weights.

3.   Of 12 coins of the same denomination, 11 weigh the same and 1 is defective (it is either heavier or lighter). Show how you can find the defective coin in three weighings on a balance scale without using weights.

4.   Of 13 coins of the same denomination, 12 weigh the same and 1 is defective (it is either heavier or lighter). Show how you can find the defective coin in three weighings on a balance scale without using weights.

5.   Of 4 coins of the same denomination, 3 weigh the same and 1 is defective (it is either heavier or lighter). Show how you can find the defective coin in two weighings on a balance scale without using weights.

6. Of 18 coins of the same denomination, 17 weigh the same and 1 is defective (it is either heavier or lighter). Show how you can find the defective coin in four weighings on a balance scale without using weights.

7. Of 24 coins of the same denomination, 23 weigh the same and 1 is defective (it is either heavier or lighter). Show how you can find the defective coin in five weighings on a balance scale without using weights.

8. How many different ways can three people be seated around a circular table?

9. How many different ways can seven people be seated around a circular table?

10. How many different ways can 11 people be seated around a circular table if two of them must always be seated next to each other?

11. Both the east-bound and the west-bound trains from a subway station ran at ten minute intervals as shown below.

East-Bound	West-Bound
12:00	12:01
12:10	12:11
12:20	12:21
12:30	12:31

A boy had a girlfriend east of this station and another girlfriend who lived west. He could never decide which girl to visit, so every day he came to the station at a random time between 12:00 and 12:30 and took the first train that came. Which girl is he likely to visit most often?

12. In many Oriental societies, people do not like to have more girls than boys, and if a woman does give birth to more girls than boys, they think something is wrong with her and blame her for the misfortune. On the basis of probability theory, decide if they are justified in putting the blame on the mother .

13. According to legend, Helen of Troy invented the game of Morra to play with Paris. In this two players game, the players face each other, each holding up a closed fist. At a given signal, they both hold up as many fingers as they wish, and announce a number from 2 to 10. A player scores a point if the total number of fingers held up by both players is equal to the number the player

announced. (a) What are the odds in favor of a player scoring a point? (b) Is this a fair game? (c) Can you think of a strategy to win this game?

14. Suppose you play Morra with both hands, that is, either player may announce any number from 2 to 20. What are the odds in favor of scoring a point?

15. (*Probability of winning at CRAPS*): Craps is played with two dice. The player rolls the dice - if the sum of the faces that turn up is 7 or 11 then the player wins. If the sum is 2, 3, or 12 then the player loses. If the sum is any number other then these, then the player rolls the dice repeatedly until he gets the same sum again, in which case he wins or until he rolls a sum of 7, in which case he loses.
   (a) What is the probability of winning on the first roll?
   (b) What is the probability of losing on the first roll?
   (c) What is the probability of winning, given that the sum on the first roll is 4?
   (d) What is the probability of winning, given that the sum on the first roll is 8?

16. The father of a budding tennis star offers a 100 dollar prize if he can win two sets in a row in a three-set series to be played with his father and his coach alternately: father-coach-father or coach-father-coach according to his choice. The son is free to choose whom he plays first. The son knows that his coach is a better player than his father. To increase his chances of winning the 100 dollar prize, who should the son choose to play first?

17. A prize on a game show is located behind one of three curtains. You choose one of the curtains. What is your probability of winning?

    After you have made your choice, the host of the game show opens one of the other curtains to show you that the prize is not there and asks you to change your choice if you wish. Should you change your choice? Decide by finding out the probability of winning after you know one of the curtains that does not have the prize behind it. [This problem is the same as the prisoner's dilemma].

18. Four passengers board a train consisting of two cars. Each passenger selects at random which car he will sit in. Find the probability of each of the following:
    E: 'there will be 2 passengers in the first car'
    F: 'there will be 2 passengers in each car'
    G: 'there will be 1 passenger in one car and 3 in the other'.

19.   Nine passengers board a train consisting of three cars. Each passenger selects at random which car he will sit in. Find the probability of each of the following:

E: 'there will be three passengers in the first car'

F: 'there will be three passengers in each car'

G: 'there will be two passengers in one car, three in another, and four in the remaining car.

20.   A woman claims that by tasting a cup of tea she can tell whether the milk was poured into the tea or the tea into the milk. To test her claim, the following experiment is designed. Eight cups of tea, four of which are prepared by pouring the milk into the tea, and four that are prepared by pouring the tea into the milk, are prepared. The order of mixing the tea and milk is to be the only known difference in the cups of tea. The lady is to be told of this fact and then the cups are to be presented to her in a random order. After tasting them she is to divide them into groups of four, according to her verdict on how they were prepared. What is the probability that

(a) She will give the correct verdict if she is only guessing?

(b) She will have 3 right and 1 wrong if she is only guessing?

(c) She will have 1 right and 3 wrong if she is only guessing?

(d) She will have 2 right and 2 wrong if she is only guessing?

21.   A tennis tournament has 8 players. The number a player draws from a hat decides his first-round run in the tournament ladder. Suppose the best player always defeats the next best and the latter always defeats all the rest. The loser of the final gets the runner-up cup. What is the probability that the second-best player wins the runner-up cup?

22.   Coupons in cereal boxes are numbered 1 to 5, and a set of one each is required for a prize. With one coupon per box, how many boxes on the average are required to make a complete set?

23.   Coupons in cereal boxes are numbered 1 to 6, and a set of one each is required for a prize. With one coupon per box, how many boxes on the average are required to make a complete set?

24.   Two baskets contain red and black balls, all alike except for color. Basket I has 5 reds and 4 blacks, and basket II has 8 reds and 7 blacks. A basket is chosen at random and you win a prize if you correctly identify the basket after

taking two balls out of the basket. After the first ball is drawn and its color noted, you can decide whether or not the ball shall be put back in the basket before taking out the second ball. How do you decide the second drawing and how do you decide on the urn?

25. In problem 24, what is the probability that you will be able to identify the basket correctly?

26. Two baskets contain red and black balls, all alike except for color. Basket I has 2 reds and 1 black, and basket II has 4 reds and 3 blacks. A basket is chosen at random and you win a prize if you correctly identify the basket after taking two balls out of the basket. After the first ball is drawn and its color noted, you can decide whether or not the ball shall be put back in the basket before taking out the second ball. How do you decide the second drawing and how do you decide on the urn?

27. In problem 26, what is the probability that you will be able to identify the basket correctly?

28. (*Fra Luca Paccioli's problem*) A and B are playing a fair game, where each one has equal chances of winning. They agreed to continue until one had won six times. The game actually stopped when A had won five times and B had won three times. How should the stakes be divided?

29. A and B are playing a fair game, where each one has equal chances of winning. They agreed to continue until one of them had won six times. The game actually stopped when A had won five times and B had won four times. How should the stakes be divided?

30. (De Mere's problem)
    (a) A single die is to be rolled 4 times. One player bets that 6 would appear at least once. What are the odds in favor of this bet?
    (b) Two dice are to be rolled 24 times. One player bets that two 6's will appear at least once. What are the odds in favor of this bet?

31. The 26 red cards from a deck of cards are to be dealt at random, 13 each to two players. What is the probability that one of the hands will get both the aces?

32. The 26 red cards from a deck of cards are to be dealt at random, 13 each to two players. What is the probability that one of the hands will get the king, the queen and the jack of hearts?

33. A pack of ten cards, numbered from 1 to 10, is shuffled and dealt into two five-card hands. What is the probability that

    (a) 9 and 10 are in the same hand?

    (b) 8, 9 and 10 are in the same hand?

    (c) Of the four highest cards, two are in one hand and the two in the other?

34. The father of a budding tennis star offers a 100 dollar prize if he can win at least two sets in a three-set series to be played with his father and his coach alternately: father-coach-father or coach-father-coach according to his choice. The son is free to choose whom he plays first. The son knows that in previous encounters he had defeated his father half the time but his coach only one third of the time. To increase his chances of winning the 100 dollar prize, whom should the son choose to play first?

    [See problem 17 of exercises 6.1. This problem is slightly different because here we do not have the requirement that the son has to win two sets in a row.]

35. A tennis player X is challenged by another player Y, who promises to pay $100 if X could beat him either in at least 3 sets out of 4 or in at least 2 sets out of 3. If X knows that Y is as good a player as he is, should he decide to play a 4 set match or a 3 set match?

36. (Rabbi Ben Ezra's problem) Jacob died and his son Reuben produced a deed duly witnessed that Jacob willed him his entire estate on his death. The son Simeon also produced a deed that his father willed to him half the estate. Levi produced a deed giving him one third, and Judah brought forth a deed giving him one quarter. How should the estate be divided? [N.L. Rabinovitch, Probability and Statistical Inferences in Ancient and Medieval Jewish Literature, Univ. of Toronto Press, Toronto, 1973].

37. The winner of a State lottery is decided by drawing a three digit number, in which each digit is decided by drawing one of ten balls each marked with a different numeral. Find the probability of each of the following:

    (a) The winning number is 092?

(b) The winning number is 092 for two consecutive days?

(c) The winning numbers of two consecutive days are the same?

   [note that 0 may appear as the first digit on a lottery number]

38.   The winner of a State's lottery is decided by drawing a four digit number, each digit of which is decided by drawing a ball from a jar of ten balls each marked with a different numeral. Find the probability of each of the following:

(a)   The winning number is 1927?

(b)   The winning number is 1927 two consecutive days? [This happened in Rhode Island on March 2nd and 3rd of 1984]

(c)   The winning numbers of two consecutive days are the same ?

39.   (Sam Loyd's puzzle) A hippopotamus, a rhinoceros and a giraffe run a race. If the odds are 2 : 1 against the hippo and 3 : 2 against the rhinoceros winning the race, what should the odds be against the giraffe?

40.   An extremely worried man finds that the probability of the plane in which he is flying being blown up by a person carrying a bomb is $1/10^6$. The probability of the plane being blown up by two persons carrying a bomb is almost 0. Therefore, to be on the safe side, he carries a bomb with him when he flies. Discuss the validity of his reasoning.

41.   (In the West there are casinos. In the street corners and market places of eastern countries, they gamble with the following three card game) A man takes three cards, one of which is a queen and the other two ordinary cards. He shuffles them very swiftly with his quick hands and puts them face down. He then asks you to put your money on the card you think is the queen. If it is the queen you get double your money and if it is not the queen, he keeps your money. Clearly, the probability that you picked the queen is 1/3 and the odds are 2:1 in his favor. Very soon, after losing your money, you realize this and decide to quit.

He then implores you "Don't go. I will give you a break. After you have put your money on a card I will turn over a card which is not the queen. Then the queen will be one of the other two and you will have an even chance to win."

(a)   Find the probability that you picked the queen given that one of the cards you have not picked is not the queen. Has the man really given you a break?

(b) What if the man says, "After you have picked your card I will let you turn over one of the other cards. If it is the queen then nobody wins, but if it is not the queen you and I will have an even chance of winning."

42. (The prisoner's dilemma) Three people, Mr., X, Mr. Y, and Mr. Z, are charged for a serious crime. Mr. X, through the warder, comes to know that one of them is sure to be hanged. He could find out from the warder, if he wanted, which one of the remaining 2 will not be hanged. He decided not to do so, thinking that the knowledge of which of the other two will not be hanged, will make him more worried. What could be his reason?

43. There are three pairs of marbles of three different colors – blue, white and red. One of each color is heavier by the same amount than the other. You have a balance scale but no weights and you have to identify the heavier marbles in two weighings.

44. Let $S = \{1, 2, 3, \ldots, n\}$. Prove each of the following:

(a) S has 2^n subsets.

(b) Every element of S belongs to 2^{n-1} subsets of S.

(c) Let S' be a collection of subsets of S such that any two members of the collection has at least one element of S in common. Then the collection S' has at most 2^{n-1} elements.

45. Find the probability of obtaining a bridge hand with 5 spades, 4 hearts, 3 diamond and 1 club.

Chapter 11

MISCELLANEOUS TOPICS

1. MONTE CARLO METHODS

Random Devices

Coins, dice, cards are random devices because we get random phenomena by tossing a coin, rolling a die, drawing a card and so on. We used random devices to explain probability theory. Now we would like to point out that random devices are very useful to imitate real world problems.

In each of the following examples, the first step is to select a suitable random device. Design an experiment that has the same number of outcomes as the real problem. Run the experiment a large number of times. There are methods to determine the number of runs needed. For the present it suffices to say that the larger the number of runs the better is the accuracy. It is also true that repeating the same experiment repeatedly may not be an interesting proposition. If there are 25 students in a class and if each student performs an experiment once, then the result is 25 runs of the experiment. Monte Carlo (see historical note for the history behind the name) is a mathematical model of a real world problem. We now give some examples of Monte Carlo method.

Examples 11.1

1. Let us assume that all of you will be married and have two children. We wish to determine how many of you will have two boys, how many are likely to have two girls, and how many are likely to have one boy and one girl.

Solution. Clearly, it will be virtually impossible to wait for you to get married and have babies. We can get a good idea of the distribution by tossing coins. Since the chance of a baby being a boy or girl is 1/2, it is like tossing a coin. If we toss a coin then there is 1/2 chance that it comes up head and 1/2 chance that it comes up tails. We designate H as boy and T as girl. So, toss a coin twice and note the outcome. Repeat it a fairly large number of times and find the frequency of each outcome. Let us say in one hundred runs of the experiment (tossing a coin twice) the distribution is as follows:

HH	HT or TH	TT
22	53	25

This means that half of you are likely to parent one boy and one girl, about a fourth will have two girls and the rest will have two boys.

2.   Coupons in cereal boxes are numbered 1 to 6, and a set of one each is required for a prize. If there is one coupon per box, determine the number of boxes that a person may need to buy to get a complete set.

Solution. Of course, we assume that the manufacturer puts equal number of coupons of each kind and distributes them at random. To answer this question one may go to the grocery store and start buying the cereal boxes -- not a very practical idea. Theoretically, it is possible that with a little bit of bad luck one may never be able to get all six. So, we turn to Monte Carlo. This time we use a die and designate each face with the corresponding coupon number and start rolling until each face turns up at least once, and note the number of rolls needed. Repeat it a large number of times and use the mean number of rolls needed as the answer.

When each student in a class of 40 did this experiment three times (this is like a run of 120) the mean was 15. So, 15 boxes is the answer.

3.   Ten members of a little league team put their hats in a box where the hats get mixed up. If each youth picks a hat at random, how many youths are likely to get his own hat?

Solution. Use ten identical cards from two suits, say A♢, 2♢, 3♢, 4♢, 5♢, 6♢, 7♢, 8♢, 9♢, 10♢ and A♠, 2♠, 3♠, 4♠, 5♠, 6♠, 7♠, 8♠, 9♠, 10♠. Designate each diamond card as one of the players and the corresponding spade card as his hat. Shuffle the cards separately and put them in two different piles. Then draw one card from each pile. If the cards match then it is a case of a player getting his own hat. Otherwise it is a case of a player getting somebody else's hat. Repeat this experiment and find the mean. [One is the answer]

4. A boy remembers all but the last figure of his girlfriend's telephone number and decides to choose the last figure at random in attempt to reach her. How many numbers will he have to dial to be able to talk to her?

Solution. Since the last digit can be any of the ten numerals 0, 1, 2, 3, 4, 5, 6, 7, 8, 9, we write each of the numerals on a piece of paper, fold it, and put it in a hat. We identify the act of dialing the last digit with the act of drawing a number from the hat. So, if the correct digit is 8, we keep drawing a number from the hat until we get the number 8. Write down the number of draws needed to get the right digit. If this is repeated a large number of times, the mean of the draws will give the answer. Remember to mix the papers in the hat before each draw and do not put back the number you draw because nobody dials a wrong number again.

Exercises 11.1

1. Design an experiment to determine what percentage of three-children families in the country have two girls and one boy.

2. Design an experiment to determine what percentage of four children families in the country have four girls.

3. Maria wishes to have five children. Design an experiment to determine which is more likely to happen a: 5-0, 4-1 or 3-2 split according to sexes.

4. Coupons in cereal boxes are numbered 1 to 5, and a set of one each is required for a prize. If there is one coupon per box, determine the number of boxes that a person may need to buy to get a complete set.

5. Nine members of a little league team put their hats in a box where the hats get mixed up. If each youth picks a hat at random, how many youths are likely to get his own hat?

6. Each box of a brand of a cereal contains one of nine cards about baseball great. Design an experiment to determine the number of boxes of the cereal a person may expect to buy to collect the entire set.

2. PASCAL'S TRIANGLE

A convenient way to obtain the numbers nCr for different values of n and r is given by Pascal's triangle, shown below. To obtain the triangle, we start with n = 0 and write 1 on top, the row n = 1 has 1, 1 and the row n = 2 has the numbers $_2C_0$, $_2C_1$, and $_2C_2$ in that order. In general, each row corresponds to the given value of n. The first and the last terms in each row are 1s since they correspond to nC_0 and nCn, respectively. Any other number on any row is the sum of the two adjacent numbers in the row just above.

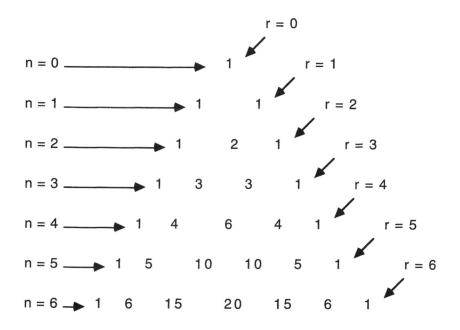

Figure 10.1

To find, say $_5C_3$, we find the number on the row n = 5 and on the diagonal r = 3, which is 10.

Binomial Theorem

The sum of two terms, say a + b, is called a binomial. The binomial theorem tells us how to write the power of a binomial expression, that is, an expression of the form

$$(a + b)^n.$$

We illustrate this by considering n = 5. The expression for

$$(a + b)^5$$

will have $5 + 1 = 6$ terms. Each of these terms contains the product of a positive exponent of 'a' and a positive exponent of 'b' such that the combined power of each term is 5. Starting with a^5 we write them as

$$a^5 + \quad a^4b + \quad a^3b^2 + \quad a^2b^3 + \quad ab^4 + \quad b^5,$$

where the power of 'a' decreases by 1 and that of 'b' increases by 1, as we move from left to right. The last term is b^5.

Then, we fill in the coefficients of each term from row n = 5 of Pascal's triangle, maintaining the order, to get

$$(a + b)^5 = a^5 + 5a^4b + 10a^3b^2 + 10\ a^2b^3 + 5\ ab^4 + b^5$$

Similarly,

$$(a + b)^2 = a^2 + 2ab + b^2$$

$$(a + b)^3 = a^3 + 3a^2b + 3ab^2 + b^3$$

In general, (Binomial theorem):

$$(a + b)^n = a^n + {}_nC_1(a^{n-1})b + {}_nC_2(a^{n-2})b^2 + \ldots + {}_nC_{n-1}(a)b^{n-1} + b^n$$

The binomial theorem follows from the fact that

$$(a + b)^n = \underbrace{(a + b)(a + b) \ldots (a + b)}_{n \text{ factors}}$$

Therefore, each term in the expansion of $(a + b)^n$ will contain the product of n combinations of a's and b's, one from each of the n factors. The first term a^n is obtained by taking only the a's from each factor of $(a + b)^n$ and there is only one way of doing it. The second term $a^{n-1}b$ is obtained by taking one b from any of the factors and one a from each of the remaining $n - 1$ factors. Since there are nC_1 ways of selecting one 'b' from n possible factors, the coefficient of $a^{n-2}b$ is nC_1. Similarly, the term $a^{n-2}b^2$ is formed nC_2 ways and so on. Hence, the theorem follows.

Examples 11.2

1. $(a + b)^8 = a^8 + {}_8C_1a^7b + {}_8C_2a^6b^2 + \ldots + {}_8C_7\,ab^7 + b^8$

$\qquad\qquad = a^8 + 8a^7b + 28a^6b^2 + \ldots + 8ab^7 + b^8$

2. $(2x - 3y)^4 = (2x)^4 + {}_4C_1\,(2x)^3(-3y) + {}_4C_2(2x)^2(-3y)^2 + {}_4C_3(2x)(-3y)^3 + (-3y)^4$

$\qquad\qquad = 2^4x^4 - 4\,(2^3x^3)3y + 6\,(2^2x^2)\,9y^2 - 4(2x)(3^3y^3) + 3^4y^4$

$\qquad\qquad = 16x^4 - 96x^3y + 216x^2y^2 - 216xy^2 + 81y^4$

Numbers of the form nCr are also referred to as *binomial coefficients* since they occur as coefficients in a binomial expression.

Exercises 11.2

1.   Putting –b in place of b in the binomial theorem, find

    (a) $(a - b)^2$ (b) $(a - b)^3$ (c) $(a - b)^5$ (d) $(a - b)^8$.

2.   Find (a) $(x + y)^2$ (b) $(x + 2y)^2$ (c) $(3x - 2y)^2$.

3.   Find

    (a) $(1 + x)^n$ (b) $(1 - x)^n$ (c) $(102)^3$ (d) $(110)^4$

    (e) 99^2 (f) $(999)^3$ (g) $(1 - \frac{1}{x})^n$

    (h) $(a + 2b)^3$ (i) $(a - 2b)^3$ (j) $(3a - b)^4$.

4.   Using the binomial theorem find: 11^2, 11^3, 11^4, 11^5.

Do you notice any similarity between your answer and the coefficients of the binomial theorem?

5.   State in words how you may find 11^n for any positive n.

6.   Write the numbers in Pascal's triangle corresponding to n = 7 and then write the expansion for $(a + b)^7$.

7.   Using the binomial theorem prove that for any positive integer n,

    (a) $nCo + nC_1 + \ldots + nCn = 2^n$

    (b) $1 \times (nC_1) + 2 \times (nC_2) + 3 \times (nC_3) + \ldots + n \times (nCn) = 2^{n-1}n$

3. HISTORICAL NOTES

The development of probability and statistics may be connected with fortune-telling, with decision making and, above all, with gambling problems. Some form of probability theory has been known and practiced since ancient times. In problem 18 of exercises 6.2, we have indicated that the fortunetellers of Asia Minor in the old days used astragali to predict good and evil events. They tossed five astragali to make a prediction - evil predictions were based on outcomes like 4, 4, 4, 6, 6 that had small probabilities $[_5C_3 \times (.35)^3 (.12)^2 = .009]$ and favorable predictions were based on outcomes like 1, 3, 3, 4, 4 that had fairly high probability $[_5C_2 \times _3C_2 \times (.37)^2(.39)^2 = .075]$. This suggests that the fortunetellers were perhaps aware of which was more probable and which was not. Decision - making problems based on expected values are found in the writings of Rabbi Ben Ezra (1092? - 1167?). The decision on the problem of Fra Luca Paccioli (1445 - 1509) was solved by using probability theory.

The modern theory of probability, however, emerged through gambling. Even though gambling has been known to mankind for thousands of years, interest in gambling reached its peak in the seventeenth century when the (gambling) aristocracy of Europe started employing professional mathematicians to determine their chances of winning in betting games. For example, the problem of the Duke of Tuscany (see 10 of exercises 3.2) was solved for the Duke by Galileo (1564 - 1642). It is because of Chevalier de Mere's problems on the gambling table that work on probability theory was begun, rather dramatically, by two French mathematicians: Blaise Pascal (1623 - 1662) and Pierre de Fermat (1601 - 1665). Pascal's friend, Chevalier de Mere, discovered some startling facts while gambling with his friends. A single die is rolled 4 times in succession. One of the gamblers would bet that 6 would appear at least once in the 4 throws, while his opponent would bet against it.

De Mere found that by tabulating the results of many games, the man betting in favor of 6 won more often. So, when they changed the game to rolling two dice 24 times, De Mere thought that the odds would still be in favor of betting on double 6 [23, fun problems]. Unfortunately, he started losing money. Since De Mere had no answer why it happened that way, he raised the question with Pascal. Pascal wrote to Fermat about this, and in their resulting correspondence, they created the theory of probability.

Pascal and Fermat, however, did not write their results. The Dutch physicist Christian Huygens, (1629 - 1695), who is credited with the effective invention of the pendulum clock, published the first book on probability theory, *De ratiocinis in ludo aleae* (On Reasoning in Games of Dice) in 1657 .

It is also known that the Italian physician, mathematician, and gambler Geronimo Cardan (1501 - 1576) had written a gambler's handbook *Liber de Ludo Aleae* (The Book on Games of Chance). However, Cardan's book did not appear in print until 1663. This is rather ironic because Cardan was a prolific writer and had published as many as 131 books in his lifetime. Cardan did not consider his book on games of chance worth publishing perhaps because he considered gambling to be a natural evil. He admits, however, "I gambled at chess more than forty years, at dice about twenty-five years, and not only every year, but - I say with shame - every day, and with the loss at once of thought, substance, and time." His book contained discussions on equiprobable measures, mathematical expectations, and the sediments of the law of large numbers. Today, most scholars agree that Cardan formulated a number of fundamental probability principles more than a century before Pascal and Fermat.

After Huygens, the mathematician Jacob Bernoulli (1654 - 1705) contributed the theorem known as the Law of Large Numbers. The law of large numbers asserts that the mean of a random sample may be made as close to the population mean as we want by taking a sufficiently large sample. Another way to state the law of large numbers is that if an experiment is repeated a large number of times, then the relative frequency of an outcome will be close to the probability of the outcome. The difference between the relative frequency of an outcome and its probability may be made as small as we wish by repeating the experiment a sufficiently large number of times. Bernoulli claimed that it took him twenty years to perfect it. His *ars conjectandi* (Art of Conjecturing), published after the author's death in 1713, was the first major work on probability theory.

The next important step was the introduction of the concept of the normal curve by De Moivre in 1733. About this time, lotteries and insurance companies started appearing everywhere. This brought some renewed interest in the theory of probability. In his work, *Annuities upon Lives,* De Moivre gave the hypothesis of equal decrements, which states that annuities can be computed on the assumption that the number of people in a given group that die is the same during each year. In 1777, George Louis Leclerc gave the first example of a geometrical probability with the so-called Buffon needle problem. Attempts were also made to compute the probability that a jury is able to arrive at a true verdict.

The probability of judgment was taken up again by Pierre-Simon Laplace (1749-1827), who, in 1812, published his *Theorie analytique des probabilities* and set the foundation of modern probability theory.

However, interest in probability theory declined during the nineteenth century even though great mathematicians like Poincare and Hilbert tried to revive some interest. In 1900, Hilbert included axiomatization of probability theory in a famous list of problems that he presented to the International Congress of Mathematicians held in Paris. Poincare also had published a set of lecture notes in probability theory.

It was, however, two great Russian mathematicians P.L. Chebyshev (1821-1894) and his student A.A. Markov (1856-1922) that brought the subject to new heights. During the last two decades, probability theory has become one of the most vigorously pursued branches of applied mathematics. It is only now that people are beginning to realize the truth of Laplace's prophetic words: "the most important questions of life are, for the most part, only problems of probability."

Statistics

The birth of statistics can be traced to the *Bills of Mortality* , a weekly publication by the parish clerks (very much like our town clerks today) in Britain, listing the births, deaths and their causes, etc. Bills of Mortality was started in 1592 but began appearing regularly only after December 29, 1603, the year of ascension of James to the throne of England. After organizing and analyzing the data in *Bills of Mortality*, John Graunt (1620 - 1674), a leisurely London businessman, published a book in 1662 with the title *Nature and Political Observations Mentioned in a following Index and made upon the Bills of Mortality*. The following are samples of John Graunt's observations and conclusions:

1. London, the metropolis of England, is perhaps a head too big for the body because this "Head grows three times as fast as the Body, that is, It doubles its people in a third part of the time".

2. The old streets are unfit for the present frequency of coaches.

3. That the vast numbers of *Beggars*, swarming up and down this City, do all live, and seem to be most of them healthy and strong.

4.   The number of males exceed the females by about a thirteenth part but since more men die violent deaths than women, the said thirteenth part difference disappears and "every woman may have a Husband, without the allowance of *Polygamy."*

5.   The Christian Religion, prohibiting *Polygamy,* is more agreeable to the *Law of Nature* than others that allow it.

6.   "I dare ensure any man at this present, well in his wits, for one in the thousand, that he shall not die a *Lunatick in Bedlam,* within these seven years because I finde not above one in about one thousand five hundred have done so."

Graunt's extraordinary observations were received with acclaim and he was made a Fellow of the Royal Society. It opened the door for ascertaining prices and annuities for lives and the first life insurance company was started in Britain in 1669. Edmund Halley(1658 - 1744), of Halley's comet, published a paper in 1692 with the title "An estimate of the Degrees of the Mortality and Funerals at the city of Breslow; with an attempt to ascertain the prices and Annuities upon Lives." In this paper, Halley not only gave a sound analysis of the calculations of annuity prices but also set the pattern for a table of mortality that is adhered to till this day. Interested students may read a text of Halley's paper and John Graunt's book in James R. Newman, *The World of Mathematics,* Vol.3, Simon and Schuster, 1966.

Graunt's observations also encouraged the gathering and study of vital statistics in continental Europe. In 1829, Lambert Quetelet (1796 - 1874), a mathematics teacher in Belgium, after planning and analyzing a census in the country, accurately predicted the crime and mortality rates from year to year. Statistical methods were introduced in the United States by Francis Galton (1822 - 1911). Vital statistics were kept irregularly in the States till 1900, when the federal government began annual death data collection from certain city and state systems. Karl Pearson (1857 - 1936) developed the concept of standard deviation and other statistical concepts. Other noteworthy contributions to the development of statistics were made by W. Weldon (1860 -1906) and Ronald Fisher (1890 - 1962).

Monte Carlo Method

The Monte Carlo method was first formalized by S.M. Ulam of Los Almos, New Mexico, in the mid 1940s. According to him, he was attempting to determine the fraction of all games of solitaire that could be completed satisfactorily to the last card. After discussing the problem with John Von Neumann, Ulam decided to have a computer play out the game a large number of times and record the results. They chose the name Monte Carlo because of the element of chance, the production of random numbers with which to play the suitable games. [Ulam, Stanislaw M. Adventures of a Mathematician, New York: Scribner's, 1976.] Monte is a gambling game played with 40-card pack. Monte Carlo is a gambling resort in Southern France. The method was used to solve problems that arose in the design of atomic reactors. Nowadays, using high speed computers, Monte Carlo method is used to solve a wide range of problems like absorption of high-energy radiation, the development of statistical tests, war games, etc.

Public Opinion Polls

The first known opinion poll was reported in *The Harrisburg Pennsylvanian* on July 4, 1824. On the basis of a straw poll taken in Willmington, Delaware, the poll showed Andrew Jackson far in the lead of other presidential aspirants. Other newspapers like the *Boston Globe* and *the New York Herald Tribune* also started conducting polls by sending reporters to street corners or by taking straw votes. These methods continued until 1916 when *the Literary Digest* conducted its first post card poll among its subscribers, and in 1920 *the Literary Digest* started the process of mailing ballots to telephone owners. Later, it included the car owners as well. Because the samples were biased, the margin of error were high in these polls. Since the predictions were on the right side, nothing controversial happened till 1936, when using the same method, *the Literary Digest* proudly predicted Landon to be the winner (57.1%) over Roosevelt (42.9%). On the election night, Roosevelt won, securing 62.5% votes. This proved a turning point in polling history and a short time later *the Literary Digest* folded.

Meanwhile, in 1933, George Gallup started experimenting with opinion polls based upon small but carefully selected samples, and correctly predicted Democratic gain in the 1934 Congressional elections. In 1935, Gallup established the American Institute of Public Opinion and started releasing its findings to sponsoring newspapers and Gallup Poll was born. Elmo Roper similarly started Roper poll based on careful samples. Public confidence in these polls started growing until 1948, when the importance of the timing factor became apparent after a big flop in predicting the winner of the presidential elections. All the opinion polls predicted a victory for Dewey and some newspapers even went ahead and published early editions declaring him to be the winner. However, Harry Truman won and many predicted that the end had come for all polls.

It is said that the media initially saw polling as competitive, as an invasion of the newsman's function and prerogatives. So, the newsmen found one more reason to hold polling back and to retard its growth. In the 1950's opinion polls did not receive as much attention as it deserved. However, it became known that in the close presidential race of 1960 between Nixon and Kennedy, both candidates relied heavily on private polls. The opinion polls for Kennedy were conducted by Louis Harris, who later in 1963 established the Harris poll, which was sponsored by newspapers across the nation. In the 70's the media became converts to polls. In fact, many of them set up their own polls.

4. BIOGRAPHICAL NOTES

BLAISE PASCAL (1623-1662) was a many-sided genius and a mathematical prodigy. He started showing mathematical talents at the age of twelve. When he was fourteen he joined his mathematician father in informal meetings of the *Mersenne' Academy* of Paris. At sixteen he proved a theorem on conics, which is now known as Pascal's theorem. At nineteen, he made the first calculating machine in history and presented it to the King of France. He then became interested in hydrostatics and made significant contributions to its development. During this period, he also proved conclusively that air has weight.

In 1654, Pascal collaborated with Fermat in laying the groundwork for the theory of probability and in giving an excellent exposition of the principle of Mathematical Induction. Fermat also got Pascal interested in the theory of numbers, but only for a brief period.

Then, on a chilly autumn night in November 1654, when he was 31 years old, Pascal experienced a religious ecstasy that caused him to abandon science and mathematics for theology. During the last eight years of his life, only once did Pascal yield to the sinful temptation to do a little mathematics - one night in 1658, when he was lying awake, tortured by a toothache, he started to think about the cycloid. The pain stopped immediately and Pascal took it as a sign from God that the study of mathematics was approved by Him.

Pascal is also famous for his dramatic wager on accepting the doctrines of the church. He argued that the doctrines of the church may be true or they may be false; it is like the flip of a coin, the odds are even. However, the payoffs are very much in favor of a bet that the church is true.

Suppose that a man rejects the church. If the church is false, he loses nothing, but if the church is true, he faces infinite suffering in Hell.

Suppose that a man accepts the church. If the church is false, he gains nothing, but if the church is true, he wins eternal bliss in heaven.

ABRAHAM DE MOIVRE (1667-1754) was a French Huguenot who settled in England after the revocation of the Edict of Nates (1685). In England, he made the acquaintance of Newton and Halley, but failed to secure a job because of his non-British origin. He supported himself as a private teacher of mathematics and spent a great deal of time with the gamblers of a London coffeehouse. In his famous book, *Doctrine of Chances*, published in 1718, he introduced numerous questions on dice games, drawing balls of various colors from a bag, and other games. In a later edition of the book, he applied probability theory to calculations of insurance and annuity premiums, dividends, etc.

In addition to the normal curve, De Moivre is also known for his theorem in trigonometry, which bears his name. Above all, De Moivre will always be remembered for the way he applied probability theory to predict the day of his death. "He discovered, when he began to fail, that it was necessary for him to sleep ten minutes longer each day then the preceding one, and figured that the day after he had slept some twenty-three and three-quarter hours, he would die in his sleep." His prediction came true.

ANSWERS

Chapter 1, Exercises 1.1

3.

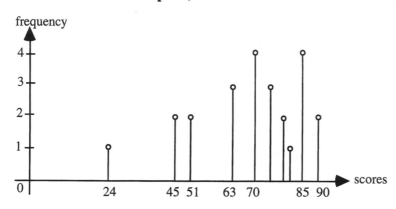

5.

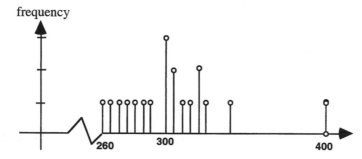

7.

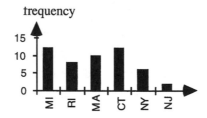

9.

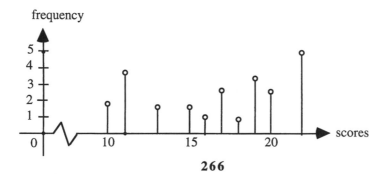

Chapter 1, Exercises 1.2

1. (a), (b)

grade	Frequency	Relative Frequency	%
A	4	4/20 = 1/5	20
B	4	4/20 = 1/5	20
C	7	7/20	35
D	3	3/20	15
F	2	2/20 = 1/10	10

3.

FINITE MATH					TOPICS IN MATH		
grade freq.	Frequency	Relative freq.			grade	Frequency	Relative
A	3	3/30			A	5	5/25
B	5	5/30			B	4	4/25
C	12	12/30			C	7	7/25
D	5	5/30			D	5	5/25
F	5	5/30			F	4	4/25

(a) Topics in Math because $\dfrac{21}{25} > \dfrac{25}{30}$

(b) Finite Math because $\dfrac{20}{30} > \dfrac{16}{25}$

(c) Topics in Math because $\dfrac{9}{25} > \dfrac{8}{30}$

(d) Finite Math because $\dfrac{22}{27} > \dfrac{16}{20}$

5. $3/32 = .09, 7/34 = .20, 15/68 = .22, 10/36 = .28, 15/53 = .28.$ On the increase

7. Construction workers, 1100.

Chapter 1, Exercises 1.3

3.

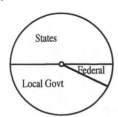

7.

1977

9.   (a) 18°   (c) 45°

11.   12.

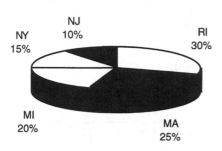

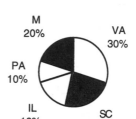

Chapter 1, Exercises 1.4

1.   1
     2   1
     3   5, 9
     4   7
     5   2, 1
     6   3, 2, 3, 3, 4, 7
     7   8, 6, 7, 0, 7, 5
     8   2, 9, 2, 8, 4, 8, 7, 6, 1
     9   3, 7, 7

     Negatively skewed

3.
     2   3, 4, 9, 7, 5, 6, 9.5
     3   2, 1, 0, 4, 4, 3, 4, 9, 9.5, 6,
          6.5, 7.5
     4   1, 7, 2, 2
     5   4, 0, 0
     6   1

     positively skewed

Chapter 1, Exercises 1.5

3.

Interval	Frequency
0 - 2	10
- 4	7
- 6	5
- 8	4
- 10	4

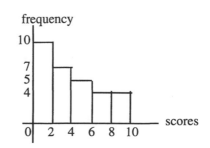

5.

Grade	Frequency
F	5
D	3
C	9
B	5
A	2

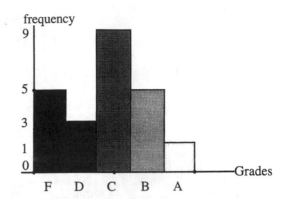

7.

Interval	Frequency
300 - 350	4
351 - 400	3
401 - 450	3
451 - 500	7
501 - 550	6
551 - 600	5
601 - 650	2

9.

Interval	Frequency
40 - 45	9
46 - 50	4
51 - 55	1
56 - 60	3
61 - 65	1
66 - 70	1
71 - 75	1

Chapter 1, Exercises 1.6

1.

Interval	f	Cum f
0 - 2	10	10
- 4	7	17
- 6	5	22
- 8	4	26
-10	4	30

5.

Interval	f	Cum f
0 - 10	0	0
11 - 20	0	0
21 - 30	2	2
31 - 40	2	4
41 - 50	3	7
51 - 60	6	13
61 - 70	4	17
71 - 80	5	22
81 - 90	5	27
91 - 100	3	30

Chapter 1, Exercises 1.7

1. $Q_1 = 25.5$,   Median $= Q_2 = 31.5$, $Q_3 = 43$

3. $Q_1 = 27$,   Median $= Q_2 = 33$, $Q_3 = 42$

5.   We have $\dfrac{n}{100} = \dfrac{400}{100} = 4$. So,

   (a) P_1 is the mean of the 4th and the 5th terms,
   (b) P_2 is the mean of the 8th and the 9th terms,
   (c) P_3 is the mean of the 12th and the 13th terms,
   (d) P_{25} is the mean of the 100th and the 1011st terms,
   (e) P_{50} is the mean of the 200th and the 201st terms
   (f) P_{75} is the mean of the 300th and the 301st terms,
   (g) P_{90} is the mean of the 360th and the 361st terms,
   (h) P_{99} is the mean of the 396th and the 397th terms.

7.   (a) 95%   (b) 95% of 80 is $.95 \times 80 = 76$.

9.   We have $\dfrac{n}{100} = \dfrac{40}{100} = .4$. So,
   (a) $5 \times .4 = 2$. So, $P_5 = 43$, the mean of the 2nd and the 3rd terms
   (b) $10 \times .4 = 4$. So, $P_{10} = 51$, the mean of the 4th and the 5th terms,
   (c) $25 \times .4 = 10$. So, $P_{25} = 63.5$, the mean of the 10th and the 11th terms,
   (d) $50 \times .4 = 20$. So, $P_{50} = 78.5$, the mean of the 20th and the 21st terms,
   (e) $75 \times .4 = 30$. So, $P_{75} = 87.5$, the mean of the 30th and the 31st terms.

Chapter 1, Exercises 1.8 (Baseball and Basketball)

1.   (a) Relative frequency of R: .146, H: .325, 2B: .066, 3B: .014, HR: .037,
                          RBI: .202, SB: .027, BB: .147, SO = .064.
   (b) .325   (c) .531   (d) 163

3.   (a) W: .068, L: .045, R: .352, ER: .326, SO: 1.10, BB: .235
   (b) 2.93.

7.   (a) 1 (.925), (2) .667, (3) .76, (4) .643, (5) .856. Yes

Chapter 2, Exercises 2.1

1. 73.3   3. 0.58   5. 4.27   11. (a) 21   (b) 91

13. (a) $x_1 + x_2 + x_3 + x_4 + x_5$ (b) $x_1^2 + x_2^2 + x_3^2 + x_4^2 + x_5^2$

   (c) $3(x_1 + x_2 + x_3 + x_4 + x_5)$ (d) $x_1 + x_2 + x_3 + x_4 + x_5 - 15$

Chapter 2, Exercises 2.2

1. (a) 7, no mode

   (b) Mean = 13.8, Mode = 14

   (c) Mean = 18.9, Mode = 11

   (d) Mean = 7.75, Mode = 7

3. Mean, 20.4

5. Mode 64, since it occurs at the center. Negatively skewed.

7.

Private School Scores		Public School Scores	
4	20	2	80
4	80, 90, 60, 60, 90, 90, 90,	3	30, 00
	80, 90, 80, 50, 90, 90, 90,	3	90, 70
	80, 50, 50, 90, 60, 70, 90	4	00, 10, 10, 30, 00
5	00, 40, 20, 30, 50, 00, 00	4	80, 80, 90, 90, 80, 70,
5	75		70, 50, 90
6	30, 00, 30, 05, 30, 10	5	00, 00, 10, 30, 30, 10,
			30, 00, 10, 00, 10
		5	80, 60, 80, 50, 90, 60,
			30, 50, 50, 60, 50
		6	00

Keith can argue that the median for the private school students is much lower than the mean, that indicates very few actually scored points above the mean (only 13 out of 35).
The median of the public school students is almost the same as the mean and more than half the students (21 out of 40) have scored more than the mean. The mean of the public school was brought down by some very low scores as can be expected in a public school (negatively skewed). Just the opposite happened in case of the private schools, that is, its mean was pushed up by 5 students who scored more than 600 points - perhaps because the private school had some bright students from traditionally privileged families.

9. Yes, symmetrical.

Chapter 2, Exercises 2.3

1. (a) 995/37 = 26.89, modal class 21 - 30.

   (b) 87.5/25 = 3.5, modal class: 6 - 10.

3. $5 \times 95 + 7 \times 85 + 12 \times 75 + 5 \times 65 + 30 = 2325/30 = 77.5$

5. -2.18 degrees Fahrenheit

7, 8. Mean for private school is 510.7 and the mean for public school is 333.1

Interval	Private School	Public School
251- 300		2
301- 350		1
351- 400		4
401- 450	4	4
451- 500	18	10
501- 550	8	12
551- 600	1	7
601- 650	6	
Modal Class	451-500	501-550

Chapter 2, Exercises 2.4

1. (a) range = 13,   Mean = 10,   M.D. = 3.3

    (b) range = 5,   Mean = 9,   M.D. = 1.6

    (c) range = 8,   Mean = 8,   M.D. = 2.6

    (d) range = 7,   Mean = 11,   M.D = 2.54

    (e) range = 8,   Mean = 6,   M.D. = 2.25

3. mean = 60, mean deviation = 11, variance = 145, s.d. = 12.04

5. (a) 50, 60, 70, 75   (b) 75   (c) None

7. (a) 6   (b) 9, 14, 12, 10, 18   (c) all of the numbers

9. Mean = 12, S.D. = 4.9. So, A, E, F, G, H, I are average, B is above average, D is excellent, C and J are below average.

11. The three players have just about the same average, but C has the smallest mean deviation, showing that C is more consistent than the other two. Select C.

13. First, find the mean and the mean deviation (or the standard deviation). If your score is within one mean (standard) deviation from the mean, then your score is average and you have a lot of company in the sense that a large number of students will have scores in that interval. Calculate the performance measure of an individual by the formula:

$$\text{Performance Measure of an individual} = \frac{\text{Deviation of the individual's score}}{\text{Mean Deviation}}.$$

A negative performance will reflect a score less than the mean and a positive performance mesure will reflect a score higher than the mean. If the performance measure is between -1 and 1, then the performance is average. If the performance measure is between 1 and 2, then the individual's performance is better than the average. If the performance measure is greater than 2, then the individual's performance is excellent. If the performance measure is less than -1, then it is below average, and so on.

Chapter 2, Exercises 2.5

1. Yes, positive

3.

X	$(x_i - \bar{x})$	$(x_i - \bar{x})^2$	Y	$(y_i - \bar{y})$	$(y_i - \bar{y})^2$	$(x_i - \bar{x})(y_i - \bar{y})$
1	-2	4	10	4	16	-8
2	-1	1	8	2	4	-2
3	0	0	6	0	0	0
4	1	1	4	-2	4	-2
5	2	4	2	-4	16	-8
Total = 15		10	30		40	-20

$$\bar{x} = \frac{15}{5} = 3, \qquad \bar{y} = \frac{30}{5} = 6 \qquad r = \frac{-20}{(\sqrt{10})\sqrt{40}} = -1$$

5.

X	$(x_i - \bar{x})$	$(x_i - \bar{x})^2$	Y	$(y_i - \bar{y})$	$(y_i - \bar{y})^2$	$(x_i - \bar{x})(y_i - \bar{y})$
5	-10	100	.15	-.35	.1225	3.5
8	-7	49	.25	-.25	.0625	1.75
10	-5	25	.35	-.15	.0225	.75
13	-2	4	.45	-.05	.0025	.10
15	0	0	.5	0	0	0
21	6	36	.6	.1	.01	.6
23	8	64	.8	.3	.09	2.4
25	10	100	.9	.4	.16	4.0
Total = 120		378	4.00		.47	13.1

$$\bar{x} = \frac{120}{8} = 15, \qquad \bar{y} = \frac{4}{8} = .5 \qquad r = \frac{13.1}{(\sqrt{378})\sqrt{.47}} = .98$$

7.

X	$(x_i - \bar{x})$	$(x_i - \bar{x})^2$	Y	$(y_i - \bar{y})$	$(y_i - \bar{y})^2$	$(x_i - \bar{x})(y_i - \bar{y})$
12	3	9	70	1	1	3
12	3	9	90	21	441	63
11	2	4	100	31	961	62
10	1	1	75	6	36	6
10	1	1	80	11	121	11
9	0	0	70	1	1	0
8	-1	1	55	-14	196	14
7	-2	4	50	-19	361	38
6	-3	9	60	-9	81	27
5	-4	16	40	-29	841	116
Total = 90		54	690		3,040	340

$$\bar{x} = 9, \qquad \bar{y} = 69 \qquad r = \frac{340}{(\sqrt{54})\sqrt{3040}} = .839.$$

9.

X	$(x_i - \bar{x})$	$(x_i - \bar{x})^2$	Y	$(y_i - \bar{y})$	$(y_i - \bar{y})^2$	$(x_i - \bar{x})(y_i - \bar{y})$
34	-10	100	46	6	36	-60
37	-7	49	47	7	49	-49
39	-5	25	44	4	16	-20
40	-4	16	20	-20	400	80
43	-1	1	45	5	25	-5
45	1	1	50	10	100	10
54	10	100	36	-4	16	-40
60	16	256	32	-8	64	-112
Total = 352		548	320		706	-196

$$\bar{x} = \frac{352}{8} = 44 \qquad \bar{y} = \frac{320}{8} = 40 \qquad r = \frac{-196}{(\sqrt{548})\sqrt{706}} = -.315.$$

Note that we took out the factor 10 from each score. This does not effect the correlation coefficient.

11.   (a) Lower.   (b) -.79   (c) Lower   (d) -.73

Chapter 3, Exercises 3.1

1.   (a) 23   (b) 7   (c)   8   (d)   5

3.   (a) 90   (b) 10   5. 2

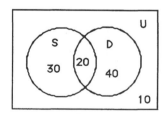

Diagram for 3

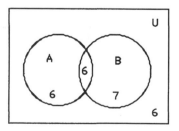

Diagram for 7

7.   (a) 13   (b) 6   (c) 0   (d) 6   (e) 6

9.   (a) 26   (b) 12   (c) 40   (d) 6   (e) 32   (f) 20.

11.   (a) 13   (b) 39   (c) 3   (d) 22   (e) 30.

Chapter 3, Exercises 3.2

1. (a) {1} (b) {a, b, c, d}

3. (a) 1 is a number, {1} is a set
   (b) ∅ has no elements but {∅} has one element. [cf. An empty can and an empty can inside another can.]

5. (a) 1, 2, 3, {1}, {1, 2} (b) Yes. (c) No. (d) Yes.
   (e) No. (f) Yes.

Chapter 3, Exercises 3.3

1. (a) {11, 12, 13, . . . , 20} (b) {1, 3, 5, . . ., 19} (c) {2, 4, 6, 8, 10}
   (d) {1, 3, 5, 7, 9} (e) {12, 14, 16, 18, 20} (e) {11, 13, 15, 17, 19}

3. (a) {0, -1, -2, -3, . . . , -10} (b) {1, -1, 2, -2, 4, -4, 5, -5, 7, -7, 8, -8, 10, -10}
   (c){3, 6, 9} (d) {1, 2, 4, 5, 7, 8, 10} (e) {0, -3, -6, -9}.
   (f) {-2, -4, -5, -7, -8, -10}

5. (a) {b, d, g, j, k, l, o, p, q, r, v, w, x, y, z} (b) {a, e, i, o, u}
   (c) {a, e, i, u} (d) {m, t, h, c, s, f, n}

7. (a) {3, 9} (b) {1, 5, 7} (c) {6} (d) {3, 9}
   (e) {2, 4, 8, 10}.

9. (a) $\{x \in U \mid x$ is a multiple of 4$\}$ (b) $\{x \in U \mid x < 11\}$
   (c) $\{x \in U \mid x$ is a multiple of 5$\}$ (d) $\{x \in U \mid x$ is a multiple of 10$\}$
   (f) $\{x \in U \mid x$ is a perfect square$\}$.

11. (a) {2, 3, 5, 7} (b) {5, 10, 15} (c) {6, 12, 18, 24, 30}
    (d) {1, 4, 9, 16, 25}

15. (b), (c) and (d) contain disjoint sets.

Chapter 3, Exercises 3.4

1. {1}, {1, 2}, {2}, ∅

3. (a) 6 (b) 2 (c) none (d) 1 (e) no

5. (a) $S \subset B$ (b) $K \subset V$ (c) $\varnothing \subset A$ (d) $B \subset A$

7. (a) $\varnothing$ (b) $\varnothing, \{a\}$ (c) $\varnothing, \{a\}, \{b\}, \{a, b\}$

9. $\{4\}, \{1, 4\}, \{2, 4\}, \{3, 4\}, \{1, 2, 4\}, \{1, 3, 4\}, \{2, 3, 4\}, \{1, 2, 3, 4\}$.

Chapter 3, Exercises 3.5

1. $A \leftrightarrow C$, $B \leftrightarrow D$, $B \leftrightarrow E$, $D \leftrightarrow E$.

3. (a) 15   (b) infinite   (c) infinite   (d) 8

    (e) infinite   (f) 1   (g) 26

5. (a) 9   (b) infinite   (c) infinite   (d) 4

    (e) 10   (f) infinite   (g) 6   (h) 6

Chapter 3, Exercises 3.6

1. (a) $\{0, 2, -2, 4, -4\} \cap \{2, 4, 6\} = \{2, 4\}$
    (b) $\{2, -2, 4, -4\} \cap \{0, 2, -2, 4, 6\} = \{2, -2, 4\}$
    (c) $\{2, 4\} \cup \{0, 2, -2\} = \{0, 2, -2, 4\}$ (d) $\{2\}$

3.   (a) 5   (b) 22

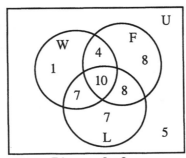

Diagram for 3

5.   (a) 100   (b) 10   (c) 15   (d) 15

Chapter 4, Exercises 4.1

1.

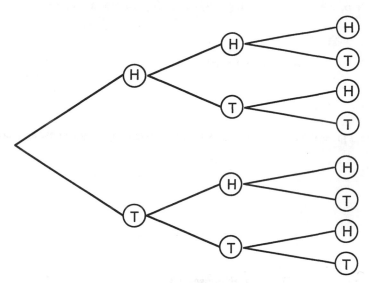

U = {HHH, HHT, HTH, HTT, THH, THT, TTH, TTT}

3.

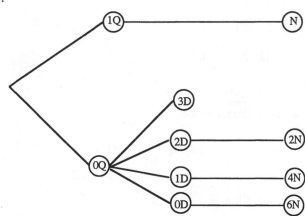

U = {QN, 3D, 2D + 2N, D + 4N, 6N}

5. U = {VSI, VSC, VSA, VChI, VChC, VChA, TSI, TSC, TSA, TChI, TChC, TChA, C1SI, C1SC, C1SA, C1ChI, C1ChC, C1ChA}

7. U = {OOA, OOL, OAO, OAA, OAL, OLO, OLA, OLL, AOO, AOA, AOL, AAO, AAL, ALO, ALA, ALL, LOO, LOA, LOL, LAO, LAA LAL, LLO, LLA}

9.   U = {H, TH, TTH, TTT}

11.   (a) {d + n, d + p, n + d, n + p, p + d, p + n}   (b) {d + n, d + p, n + p}

Chapter 4, Exercises 4.2

1. $2^5 = 32$

3. $5 \times 4 \times 3 \times 2 \times 1, 2 \times 4 \times 3 \times 2 \times 1$

5. 8×10^6 (Since 0 gives the operator and '1' is for long distance), create a new area code.

7.   (a) 52   (b) 52×51

9.   (a) 24   (b)   10

11. $13 \times 2 + 39 \times 51$

Chapter 4, Exercises 4.3

1.   (a) 720   (b) 4   (c) 3628800   (d) 210   (e) 5040
      (f) 5040   (g) 1   (h) $_{10}C_3 = 120$

3.   6!   5. $_{15}P_{11}$ 7. $_7P_3$

9. 4×4 11. $2 \times 4! \times 4!$

Chapter 4, Exercises 4.4

1.   (a) 10   (b) 21   (c) 495   (d) 1   (e) 1   (f) 50
      (g) 19,900   (h) 8

3.   (a) $_{25}C_3 \times _{25}C_4$ (b) $_{25}C_3 \times _{22}C_4$

5.   (a) $_{15}C_2 = 105$ (b) $2 \times _{15}C_2 = 210$

7.   (a) 2^5 (b) 10

9.   (a) $_{10}C_2 \times _{12}C_3$ (b) $_{10}C_3 \times _{12}C_3 \times _7C_2 \times _9C_2$

11. $_{10}C_8 = _{10}C_2 = 45$ 13. $_{10}C_5$

15. $_{10}C_8 + _{10}C_9 + _{10}C_{10}$ 17. $_{10}P_2$

Chapter 5, Exercises 5.1

1. (a) U = {HH, HT, TH, TT}   (b) E = {TT}

3. (a) U = {HHH, HHT, HTH, HTT, THH, THT, TTH, TTT}
   (b) {HHT, HTH, THH}
   (c) {HHH, HHT, HTH, HTT, THH, THT, TTH}

5. (a) {(4, 6), (5, 5), (6, 4)}
   (b) {(1, 6), (2, 5), (3, 4), (4, 3), (5, 2), (6, 1)}
   (c) {(4, 6), (5, 5), (5, 6), (6, 4), (6, 5), (6, 6)}
   (d) {(4, 1), (4, 2), (4, 3), (4, 4), (4, 5), (4, 6), (1, 4), (2, 4), (3, 4), (5, 4), (6, 4)}

7. (a) {BB, WB, BW}   (b) {WB, BW, BB}   (c) {WB, BW}   (d) ∅

Chapter 5, Exercises 5.2

1. 1/3, 4/9

3. (a) 2/7   (b) 0   (c) 3/7

5. 15/21 = 5/7

7. U = {BB, BG, GB, GG}   (a) 1/2   (b) 1/2   (c) 1/4

9. (a) 3/8   (b) 1/8

11. Pr[2] = Pr[12] = 1/36, Pr[3] = Pr[11] = 2/36, Pr[4] = Pr[10] = 3/36,
    Pr[5] = Pr[9] = 4/36, Pr[6] = Pr[8] = 5/36, Pr[7] = 6/36

13. There are 16 possibilities, of which 6 are for 2-2 split and 8 are for 3-1 split, hence
    the latter has a higher probability.

Chapter 5, Exercises 5.3

1. Pr[E] = 1/13, Pr[F] = 3/13

3. (a) 52   (b) 39   (c) 3/4   (d) 12   (e) 3/13
   (f) The set of face cards that are not diamond   (g) 9   (h) 9/52

5. (a) 5/36   (b) 11/36

7. (a) $\dfrac{1}{36}$ (b) $\dfrac{1}{3} \times \dfrac{1}{3} = \dfrac{1}{9}$

9. (a) 52×51 (b) 13×12 (c) $1/17$ (d) 4×3

    (e) $1/221$ (f) $\varnothing$ (g) 0 (h) 0

11. (a) 8 (b) 3 (c) $4/16$ (d) 1 (e) $1/16$ (f) 15

    (g) $15/16$

13. $\Pr[E] = \dfrac{{}_8C_4}{2^8} = \dfrac{70}{256} = \dfrac{35}{128}$, $\Pr[F] = \dfrac{{}_8C_6 + {}_8C_7 + {}_8C_8}{256} = \dfrac{37}{256}$

15. $\dfrac{12 \times 11 \times 10}{52 \times 51 \times 50} = \dfrac{11}{1105}$

17. (a) $U = \{(W_1W_2,\ R_1R_2),\ (W_1R_1,\ W_2R_2),\ (W_1R_2,\ W_2R_1),\ (W_2R_1,\ W_1R_2),$
    $(W_2R_2,\ W_1R_1),\ (R_1R_2,\ W_1W_2)\}$.

    (b) $2/6 = 1/3$.

19. (a) $1/6$ (b) $10/36 = 5/18$

21. The following are the possible values for the sum and the corresponding
    probabilities:

3, 18	4, 17	5, 16	6, 15	7, 14	8, 13	9, 12	10, 11
$\dfrac{1}{216}$	$\dfrac{3}{216}$	$\dfrac{6}{216}$	$\dfrac{10}{216}$	$\dfrac{15}{216}$	$\dfrac{21}{216}$	$\dfrac{25}{216}$	$\dfrac{27}{216}$

22. The sum of 9 may be obtained as $1 + 2 + 6$ in $3! = 6$ ways, as $1 + 3 + 5$ in 6
    ways, as $2 + 3 + 4$ in 6 ways, as $2 + 2 + 5$ in 3 ways, as $4 + 4 + 1$ in 3 ways and
    as $3 + 3 + 3$ in 1 way giving a total of 25 possibilities. But the sum of 10 can be
    obtained as $1 + 3 + 6$ in 6 ways, as $1 + 4 + 5$ in 6 ways, as $2 + 3 + 5$ in 6 ways,
    as $2 + 2 + 6$ in 3 ways, as $3 + 3 + 4$ in 3 ways and as $4 + 4 + 4$ in 3 ways giving
    a total of 27 possible ways. Hence the result follows.

Chapter 5, Exercises 5.4

1. (a) $1/4$ (b) $3/4$ (c) $1/13$ (d) $12/13$

    (e) $1/52$ (f) $4/13$ (g) $3/52$ (h) $10/13$

3. (a) $4/13$ (b) $9/13$ (c) $17/26$ (d) $11/13$

5. (a) $1/2$ (b) $1/2$ (c) 0 (d) $2/3$

7. (a) $F = \{(3, 1),(1, 3), (3, 2), (2, 3), (3, 3), (3, 4),(4, 3), (3, 5), (5, 3), (3, 6),(6, 3)\}$
    (b) $11/36$ (c) $25/36$ (d) $30/36$ (e) $2/36$
    (f) $15/36$ (g) $21/36$ (h) $34/36$ [Note $\overline{E} \cup \overline{F} = \overline{E \cap F}$]

9. (a) 3/5 (b) 4/5 (c) 1/5

11. $\Pr[E] = \dfrac{255}{256}$, $\quad$ $\Pr[F] = \dfrac{247}{256}$

13. $\dfrac{19}{34}$

15. (a) 1/17 (b) 3/13 (c) $\dfrac{3}{221}$ (d) $\dfrac{61}{221}$

17. (a) 1/4 (b) 5/221 (c) $\dfrac{181}{884}$

Chapter 5, Exercises 5.5

1. (a) 1 : 3 (b) 1 : 3 (c) 1 : 5
    (d) 1 : 16 (e) 1 : 11

3. (a) Expected win $= 4 \times \dfrac{1}{4} = 1$

    (b) Expected loss $= 2 \times \dfrac{3}{4} = \dfrac{3}{2}$

    (c) No because expected loss is not equal to expected win.

5. U = {HH, HT, TH, TT}. Expected win $= 1 \times \dfrac{3}{4}$; Expected loss $= 1 \times \dfrac{1}{4}$. No.

7. 3 : 1, pay 1 dollar if it wins and receive 3 dollars if it loses.

9. (a) 3/8, (b) 5/8

11. (a) Expected win $= 36 \times \dfrac{1}{38}$, Expected loss $= 1 \times \dfrac{37}{38}$

    (b) Expected win $= 1 \times \dfrac{18}{38}$, Expected loss $= 1 \times \dfrac{20}{38}$

13. 4 girls, 3 blondes

15. Pr[Bush] = 4/9, Pr[Carter] = 1/6. Since there are only three runners,
    Pr[Bush] + Pr[Carter] + Pr[Ford] = 1. Therefore, Pr[Ford] = 7/18. So, the odds in
    favor of Ford winning are 7 : 18.

Chapter 5, Exercises 5.6

1. (a) $\dfrac{6}{17}$ (b) $\dfrac{2}{17}$ (c) $\dfrac{4}{17}$ (d) $\dfrac{1}{3}$

3. (a) $\dfrac{48}{51}$ (b) $\dfrac{12}{51}$ (c) $\dfrac{12}{48}$ (d) $\dfrac{12}{13}$

5. E: It is a spade, F: It is not a face card; 1/4, independent

7. E: It is a heart, F: It is not a spade; 1/3, not independent

9. (a) E: The sum is 8, F: One is a 4; 1/11   (b) 1/6

11. E: A multiple of 3 turns up, F: An even number turns up; 1/3, independent

13. E: 'Heads at least twice', F: 'First toss is a head'; 3/4, not independent

15. E: 'Both cards are hearts', F: 'Both cards are red'.
    (a) 4/17   (b) 6/25   (c) 11/188

17. E: The correct answer, F: More true than false etc.
    (a) 1/32   (b) 1/16   (c) 1/8
    (d) 1/4   (e) 1

19. The wife is wrong. The probability of the next child being a girl is still 1/2. The sex of a baby is independent of what happened before.

21. E: It is a vowel, F: It is a letter in 'establishment'. $Pr[E \mid F] = 11/26$

Chapter 6, Exercises 6.1

1. 1/4

3. (a) 1/15

    (b) 2/5

    (c) 1/5

    (d) 3/5

5. (a) 1/9   (b) 1/3

7. 3/4   9. 3/4

11. Room A has probability 19/36.

13. (a) 5/9

    (b) 4/9

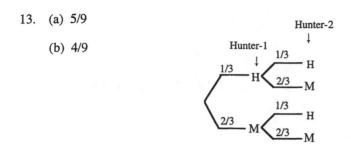

15. 1/2.

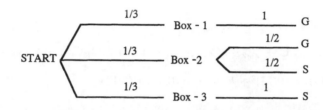

17. The probability of winning a coach-father-coach series is (1/6) + (1/9) = 5/18 (see diagram- note that he has to win two sets in a row). Similarly, the probability of winning a father-coach-father series is (make diagram and check) is (1/6) + (1/18) = 1/4. Therefore, he should play the coach first. Another way to look at this problem is to note that to win two sets in a row he will have to win the second match, no matter whom he plays first. In a father-coach-father series, his probability of winning the second set is 1/3; whereas the probability of winning the second set in a coach-father-coach series is 1/2. So, he should play the coach first.

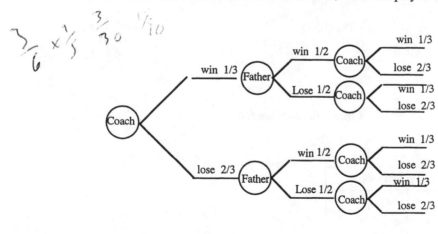

19. (a) 16   (b) Yes, 1/16   (c) 4

    (d) 6   (e) 3/8

Chapter 6, Exercises 6.2

1. $_5C_3(1/6)^3(5/6)^2$

3.   (a) $_7C_4(.4)^4(.6)^3$ (b) $_{10}C_7(.3)^7(.7)^3$ (c) $21(.4)^5(.6)^2 + 7(.4)^6(.6) + (.4)^7$
   (d) $_{10}C_8(.7)^8(.3)^2 + _{10}C_9(.7)^9(.3) + (.7)^{10}$

5. $1 - _5C_0(_1/4)^0(3/4)^4 = 1 - (81/256)$

7. $_{10}C_8(.6)^8(.4)^2$

9. $(1/2)^{10}[_{10}C_7 + _{10}C_8 + _{10}C_9 + _{10}C_{10}]$

11.   (a) $(1/2)^{12}[_{12}C_0 + _{12}C_1 + _{12}C_2 + _{12}C_3]$ (b) $1 - (1/2)^{12}[_{12}C_0 + _{12}C_1 + _{12}C_2 + _{12}C_3]$

13. $3/13, \quad 21(3/13)^5 \times (10/13)^2 + 7(3/13)^6 \times (10/13) + (3/13)^7$

15.   (a) $_{12}C_3(1/5)^3(4/5)^9$ (b) $(4/5)^{12}$ (c), (d), (e) $1 - (4/5)^{12}$ (f) $(4/5)^{12}$

17. $1 - (2/3)^{10}$

Chapter 6, Exercises 6.3

1.   (a) .0439   (b) .0547

3.   (a) .0439   (b) .0547   (c) .999   (d) .0547

5.   .2639

Chapter 6, Exercises 6.4

1.   (a)   (b) 2

n	Probability
0	1/16
1	1/4
2	3/8
3	1/4
4	1/16

3. (a)   (b) .5

n	Probability
0	125/216
1	75/216
2	15/216
3	1/216

5. (a) Pr[both hearts] $= \dfrac{1}{17}$. So, $34 \times \dfrac{1}{17} = 2$ (b) $2 \times 4 = 8$

7. 8   9. One third of the total number of questions

11. 1,100   13. $10(1 - (.9)^{16}) = 8$ (approx.)

Chapter 6, Exercises 6.5

1. 15

3. $\sqrt{4500 \times \dfrac{1}{6} \times \dfrac{5}{6}} = \sqrt{\dfrac{5 \times 9 \times 10 \times 10 \times 5}{6 \times 6}} = \dfrac{5 \times 3 \times 10}{6} = 25$

5. (a) 40   (b) 48

7   Mean $= 960$, s.d. $= \sqrt{2400 \times .4 \times .6} = \sqrt{24 \times 4 \times 6} = 24$. So $\dfrac{996 - 960}{24} = 1.5$

9. Mean $= 5,000$, s.d. $= 50$. So, deviation in standard unit is 3.

11. $p = .8$, Mean $= 900 \times .8 = 720$, s.d. $= 12$. No, not necessarily.

13. Yes

15. $p = .1$, Mean $= 40,000 \times .1 = 4,000$, s.d. $= 60$. Yes, the machine was not repaired properly.

17. (a) 0.0011, 1.8698, yes   (b) 0.3101, 9.744, no

Chapter 7, Exercises 7.1

1. (a) The graph will drift to the right and will be more spread out.

    (b) 1   (c) 6

3. (a) 1   (b) 4   (c) No

     (d) The graph for $p = .4$ will be like the graph for $p = .6$ drawn backwards.

5. (a) 2   (b) 1   (c) 2   (d) No.

Chapter 7, Exercises 7.2

1. (a) 1   (b) -1.5   (c) 2.5   (d) -2.5

3. 2 units less than the mean, $z = -2$

7. $p = .75$, Mean $= 900$, s.d. $= 15$, deviation in standard units $= 2.\overline{6}$

Chapter 7, Exercises 7.3

1. (a) 0.8804   (b) 0.86333   (c) 0.34325   (d) 0.3849   (e) 0.3050

3. (a) 0.0228   (b) 0.9821   (c) 0.8413   (d) 0.5

5. (a) 1.1   (b) -1   (c) 1.2   (d) -1.2

Chapter 7, Exercises 7.4

1. Mean $= 800$, s.d. $= 20$.

     (a) $x > 790$ gives $z > \dfrac{790 - 800}{20} = -.5$. Find the area under the normal curve for $z > -.5$. Answer: $.1915 + .5 = .6915$.

     (b) $x < 830$ gives $z < \dfrac{830 - 800}{20} = 1.5$. Find area under the normal curve for $z < 1.5$. Answer $.4332 + .5 = .9332$.

     (c) x between 790 and 830 means $790 < x < 830$. So, $\dfrac{790 - 800}{20} < z < \dfrac{830 - 800}{20}$ or $-.5 < z < 1.5$. Hence the answer is $.1915 + .4332 = .6247$.

     (d) $785 < x < 825$ gives $-7.5 < z < 1.25$. Rounding to the tenth $-.8 < z < 1.3$. Hence the answer is $.2881 + .4032 = .6913$.

3. Mean $= 3,000$, s.d. $= 50$. So, z must be between $\dfrac{2920 - 3000}{50} = -1.6$

and $\dfrac{3020 - 3000}{50} = 0.4$. Hence the answer is $.4452 + .1554 = .6006$.

5. Mean = 450, s.d. = 15. So, z must be less than $\dfrac{432 - 450}{15} = -1.2$. Hence the answer is $.5 - .3849 = .1151$.

7. Mean = 50, s.d. = 5. So, z must be greater than $\dfrac{60 - 50}{5} = 2$. Hence the answer is $.5 - .4772 = .0228$.

9. (a) 0.7745   (b) 0.0228

Chapter 7, Exercises 7.5

1. p = .5, Mean = 200, s.d. = $\sqrt{400 \times .5 \times .5} = 10$. In this case, we need to determine c' such that the area under the normal curve for z < c' is .9. Hence c' should be such that the area between 0 and c' is $0.9 - 5 = 0.4$. This gives c' = 1.3. Then c should be such that
$$\frac{c - 200}{10} = 1.3, \text{ that is } c - 200 = 1.3 \times 10 = 13 \text{ or}$$
$$c = 200 + 13 = 213.$$
You would predict that the coin would come heads fewer than 213 times.

3. p = $\dfrac{1}{6}$, Mean = 3,000, s.d. = $\sqrt{18000 \times \dfrac{1}{6} \times \dfrac{5}{6}} = 50$. In this case, we need to determine c' such that the area under the normal curve for z < c' is .99. Hence c' should be such that the area between 0 and c' is $0.99 - 5 = 0.49$. This gives c' = 2.4. Then c should be such that
$$\frac{c - 3000}{50} = 2.4, \text{ that is } c - 3000 = 2.4 \times 50 = 120 \text{ or}$$
$$c = 3000 + 120 = 3120.$$

5. 475

7. p = .25, Mean = 75, s.d. = $\sqrt{300 \times .25 \times .75} = 7.5$. In this case, we need to determine c' such that the area under the normal curve for z < c' is .9. Hence c' should be such that the area between 0 and c' is $0.9 - 5 = 0.4$. This gives c' = 1.3. Then x should be such that
$$\frac{c - 75}{7.5} = 1.3, \text{ that is } c - 75 = 1.3 \times 7.5 = 9.75 \text{ or}$$
$$c = 75 + 9.75 = 84.75 = 85 \text{ (approx.)}$$

Chapter 8, Exercises 8.1

1. (a) - 94 cents

    (b) - 94 dollars. No, it does not significantly increase your chances of winning

    (c) 940,000 dollars

    (d) It maybe because of any of the following reasons:
         (1) The expectation rises.
         (2) Everybody sees it as a chance of making a lot of money the easy way.
         (3) People are crazy to become rich overnight.
         (4) Some people need money desparately.

3. -1/2 dimes, no.

5. -12/36, not fair.

7. (a) 0   (b) 0
    (c) Favors Jim because John's or Joe's expected win = 2 × (1/4) - 1 × (3/4) = -(1/4)
         but Jim's expected win is 2 × (1/2) - 1 × (1/2) = 1/2.

9. 0.88, 8.8

11. (a) $2 \times \dfrac{1}{216} + 1 \times \dfrac{15}{216} - 1 \times \dfrac{125}{216} = -\dfrac{1}{2}$ (b) $6 \times \dfrac{1}{2} = 3$

Chapter 8, Exercises 8.2

1.

u	HH	HT	TH	TT
X(u)	2	1	1	0

r	0	1	2
Pr[X = r]	1/4	2/4	1/4

Expected value of X = 1.

3.

u	H	TH	TTH	TTTH	TTTT
X(u)	1	2	3	4	4

r	1	2	3	4
Pr[X = r]	1/2	1/4	1/8	1/8

Expected number of tosses = $1 \times \dfrac{1}{2} + 2 \times \dfrac{1}{4} + 3 \times \dfrac{1}{8} + 4 \times (\dfrac{1}{16} + \dfrac{1}{16}) = \dfrac{15}{8}$.

5.

r	-1	-.5	1
Pr[X = r]	1/8	3/8	1/8

7.

r	1	4	9	16	25	36
Pr[X = r]	1/6	1/6	1/6	1/6	1/6	1/6

Expected value $= \frac{1}{6}(1 + 4 + 9 + 16 + 25 + 36) = \frac{91}{6} = 15.\,17.$

9.

r	0	1	2	3	4	5
Pr[X = r]	1/32	5/32	10/32	10/32	5/32	1/32

Expected value of X = 2.5.

Chapter 8, Exercises 8.3

1. $(1 + 2 + 3 + 4 + 5) \times (1/5) = 3$

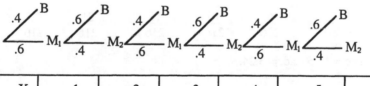

3.

X	1	2	3	4	5	6
p	.4	.36	.096	.0864	.02304	.03456

Expected number is 2.076.

5.   Let X = Number of hearts. Then

X	outcomes	Probability
1	one heart	$\frac{13 \times 39 \times 2}{52 \times 51} = \frac{13}{34}$
2	two hearts	$\frac{13 \times 12}{52 \times 51} = \frac{1}{17}$
0	no hearts	$\frac{19}{34}$

Expected value of X = .5.

7.   The daughter because her expected allowance is $15. 1 whereas that of the son is $15.

Chapter 8, Exercises 8.4

1.   6   3.   9

5.   13   7. $\frac{13}{3}$ or at least 4

9.   5   10. $\frac{5}{2}$ or at least 3.

Chapter 8, Exercises 8.5

1.   variance = .6875, s.d. = .8291

3.   mean = 7, variance = 5.83, s.d. = 2.4

5.   mean = 0, variance = 6.8, s.d. = 2.61

7.   The following are the possible values for the sum and the corresponding probabilities:

3, 18	4, 17	5, 16	6, 15	7, 14	8, 13	9, 12	10, 11
$\frac{1}{216}$	$\frac{3}{216}$	$\frac{6}{216}$	$\frac{10}{216}$	$\frac{15}{216}$	$\frac{21}{216}$	$\frac{25}{216}$	$\frac{27}{216}$

variance = 10.5   s.d. = 3.24.

Chapter 8, Exercises 8.6

1.   (a) at least 3/4   (b) at least 8/9   (c) at least 21/25
      (d) at least 5/9   (e) at least 7/16

3.   3/4

5.   10/49, since Mean = 14, $k^2 = 49/39$.

7.   5/16, since Mean = 10, $k^2 = 16/11$.

8.　　(a)　$k = \dfrac{1060 - 1000}{50} = \dfrac{6}{5}$ so the probability is at least $1 - \dfrac{25}{36} = \dfrac{11}{36}.$ The

required number of light bulb is at least $100 \times \dfrac{11}{36} = 30.55$

(b) $100 \times \dfrac{3}{4} = 75$　　(c) $100 \times \dfrac{21}{25} = 84$

(d) $100 \times \dfrac{8}{9} = 88.8$　　(e) $100 \times \dfrac{15}{16} = 93.75$

9.　　(b) 55.5 %　　(e) 80%

Chapter 9, Exercises 9.1

1.　No. Everybody in the class did not have the same chance of being selected. A random sample may be obtained by the same procedure as the one indicated in example 3.

3.　No. It will be very difficult, if not impossible, to take a random sample of sixteen men without limiting ourselves to a small community.

5.　Write all the two digit numbers on slips of paper and draw from a hat.

7.　Yes.

11.　Three digits are needed to label 235 houses (say 001 to 235). Read three digit groups from the random digits table. Discard digits that result in a number higher than 235. Thus starting at the beginning of the 7th row we can pick house numbered 075, 112, 042, 204.

Chapter 9, Exercises 9.2

1.　Senator Gore is wrong because time and again it has been established that the process works both on theory and in practice. It is like determining the quality of a soup by tasting a spoonful or two. Even though people are not like water or like the ingredients of a soup, even though as individuals people are widely different, in the mass people do conform to certain patterns of behavior. The behavior of the sample, no matter how small, will reflect the behavior of the population provided the sample has been carefully selected to represent a broad based cross-section. The important point is, as Gallup puts it, 'the sampling process must be conducted with great care to make certain that all major variations or departures from the norm are embraced. ..... Since some differences that exist may be unknown to the researcher, his best procedure to be sure of representativeness is to select samples from the population by a chance or random process.'

2. [It is estimated that 85% of the adult Americans have telephones nowadays. According to Burns W. Roper, the chairman of the Roper Center at the University of Connecticut, Storrs, 'The fact is that non-telephone households are disproportionately poor, black and rural. It was also my experience while at Gallup that, compared with our personal interview surveys, our telephone surveys tended to overrepresent those who had not gone beyond grade school. Undoubtedly as a result of these socio-economic differences, our telephone surveys also tended to be more Republican. To control for this source of bias, we used standard statistical controls, namely, "weighing" the obtained sample.']

3. $\dfrac{1500}{210000000} = \dfrac{1}{140000}$

4. See answer to 1 and 3. The odds that the correspondent will meet a person polled is 1: 139999.

5. These are not sent to random samples. A large number of persons do not send back their response. In addition, this is an asseement of opinion with own interests in mind. Therefore, they are likely to be biased in designing and in the types of questions that are asked.

7. 64% is close to 60 so the margin of error is 4 percentage points.

9. .028 or .03 (approximately), [82.97, 83.03]

Chapter 9, Exercises 9.3

1. $p = 20\% = .2$, $1- p = .8$, $n = 36$. So, standard deviation = $\sqrt{\dfrac{.2 \times .8}{36}} = .066$,

3. Mean = 60, s.d. = $\sqrt{\dfrac{.4 \times .6}{96}} = .05$

5. Mean = 20, s. d. = $\sqrt{\dfrac{.2 \times .8}{100}} = .04$

Chapter 9, Exercises 9.4

1. (a) .1   (b) .033   (c) .029   (d) .026   (e) .026

2. $z = 1.6$ gives margin of error $= \dfrac{1.6}{2\sqrt{n}} = \dfrac{.8}{\sqrt{n}}$.

    (a) .08   (b) $\dfrac{.8}{\sqrt{800}} = \dfrac{.8}{20\sqrt{2}} = \dfrac{.04}{\sqrt{2}} = .028$

(c) .023   (d) .021   (e) .02

3. (a) .428   (b) [.428 - .1428, .428 + .1428] = [.272, .5708]

5. (a) size of the sample.   (b) No effect.   (c) smaller   (d) larger.

7. (a) .75   (b) z = 1.3, margin of error = .08, Confidence interval = [.67, .83].

Chapter 9, Exercises 9.5

1. [0.20, 0.30]   3.   [0.07, 0.33]

5. Yes, since the 95 percent confidence interval for the probability of success is [0.656, 0.906]

7. [0.135, 0.265], smaller.

Chapter 9, Exercises 9.6

1. [800, 4000]   3.   [714, 10,000]

Chapter 9, Exercises 9.7

1. $n = 152$, $x = 100$

2. $.5 - A(z) = 1 - .95 = .05$. So, $A(z) = .5 - .05 = .45$. So, $z = 1.6$.

$$r = \frac{\sqrt{.6 \times .4} + \sqrt{.7 \times .3}}{.1} = 9.48155518.$$ So, $n > z^2 r^2$ gives $n > 230.143$.

Taking $n = 231$, we get $x > 231 \times .6 + (1.6) \sqrt{231 \times .6 \times .4}$ or $x > 150.51$.

So, give a test with 231 questions. The student's claim is accepted if he or she gets more than 150 questions correct.

3. $n = 314$, $x = 1$. Vaccinate 314. If no one gets the 'flu then the manufacturer's claim is justified.

4. $.5 - A(z) = 1 - .99 = .01$. So, $A(z) = .5 - .01 = .49$. So, $z = 2.4$.

$$r = \frac{\sqrt{.0001 \times .9999} + \sqrt{.01 \times .99}}{.0099} = 11.06042866 . \text{ So, } n > z^2 r^2 \quad \text{gives}$$

$n > 704.639$. So, taking $n = 705$, we get

$x > 705 \times .0001 + (2.4) \sqrt{705 \times .0001 \times .9999}$ or $x > .708$.

So, vaccinate 705. If no one gets the 'flu then the manufacturer's claim is justified.

5. $n = 2293, \ x = 1090$

Exercises 10.1

1, 2   To find how to distribute the coins in each weighing, find the set of all possible outcomes. This may be done in the following way. In a weighing three things may happen: (1) the left pan goes down ($\downarrow$) and so, it contains the heavier coin, (2) the left pan goes up ($\uparrow$), the right pan contains the heavier coin, and (3) the pan balances (=), neither pan contains the defective coin. The set of outcomes may then be given as

$$U = \{\downarrow\downarrow, \downarrow\uparrow, \downarrow=, \uparrow\downarrow, \uparrow\uparrow, \uparrow=, =\downarrow, =\uparrow, ==\}.$$

The idea is to associate with each of these ouccomes one coin, say

$\downarrow\downarrow$ (1) , $\downarrow\uparrow$ (2) , $\downarrow=$ (3), $\uparrow\downarrow$ (4) , $\uparrow\uparrow$ (5) ,

$\uparrow=$ (6) , $=\downarrow$ (7) , $=\uparrow$ (8) , $==$ (9).

Table 1

So, we label each coin with numbers 1 - 9 and then arrange the weighings so that each outcome will give the corresponding coin as the the heavier one. For example to know that the coin (9) is heavier, it should be the only coin not used in either weighing. To know that coin 4 is the heavier one, it should be the only coin to be used on the right pan in the first weighing and on the left pan in the second weighing.   Continuing this way, the weighings should be as follows:

	Left pan	Right pan
First weighing:	1, 2, 3	4, 5, 6
Second weighing:	1, 8, 4	5, 2, 7
coin 9 not used.		

Then observing the outcome, we can find out the defective coin from table 1.

3,4   These problems are different from 1 and 2 because this time we do not know whether the coin is heavier. Therefore, the outcomes $\downarrow\downarrow\downarrow$ and $\uparrow\uparrow\uparrow$ are the same and neither of them give us any information. Therefore, we discard these as possible outcomes. Using the same notation as used in 1, 2 we get the following set of outcomes and associating the coins as shown in table 2, we make the weighings as shown below.

↓↓↑ (1),   ↓↓= (2),   ↓↑↓ (3),   ↓↑↑ (4),   ↓↑ = (5),
↓=↓ (6),   ↓=↑ (7),   ↓= = (8),   =↓↓ (9),   =↓↑ (10)
=↓= (11),   ==↓ (12),   = = = (13).

	Left pan	Right pan
First weighing:	1, 3, 7, 8	2, 4, 5, 6
Second weighing:	1, 4, 5, 11	2, 3, 9, 10
Third weighing:	3, 4, 10, 12	1, 6, 7, 9

Table 2

In this case, coin 13 is not used; if coin 13 is the defective one, we cannot determine if it is heavier or lighter.

8.   In this case the answer is not 3! since the permutations a b c, c a b, and b c a give the same arrangement around a circular table. The way to answer this is to fix the position of one of the three and find the number of permutations of the remaining 2, which is 2! or 2 (see diagram below).

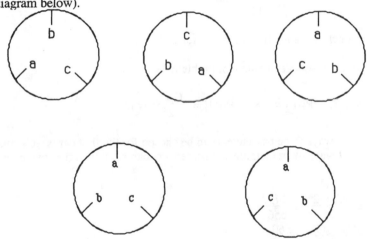

9.   6!

10. 2 × (9!)

11. The girl on the east side, since the waiting period between a westbound and an eastbound train is 9 minutes, whereas the waiting period between an eastbound and a westbound train is only 1 minute.

13. A player knows how many fingers he or she is going to hold up. So, there are only five different possibilities for the sum of the number of fingers to be put up by the two players. Each player will announce one of these five possibilities. Hence the probability of getting the correct number is 1/5. So, (a) [1: 4], (b) Yes, this is fair game because the odds of winning for each palyer is the same. (c) No.

14. [1: 9]

15. Let Pr[n] represent Pr[the sum of the two faces is n]. Then, we get $Pr[2] = \frac{1}{36}$, $Pr[3] = \frac{2}{36}$, $Pr[4] = \frac{3}{36}$, $Pr[5] = \frac{4}{36}$, $Pr[6] = \frac{5}{36}$, $Pr[7] = \frac{6}{36}$, $Pr[8] = \frac{5}{36}$, $Pr[9] = \frac{4}{36}$, $Pr[10] = \frac{3}{36}$, $Pr[11] = \frac{2}{36}$, $Pr[12] = \frac{1}{36}$. Therefore,

(a) $Pr[7] + Pr[11] = \frac{8}{36}$ (b) $Pr[2] + Pr[3] + Pr[12] = \frac{4}{36}$

(c) There are 3 ways of winning (getting sum 4) and there are 6 ways of losing (getting sum 7). So, the probability of winning is $\frac{3}{3+6} = \frac{1}{3}$

(d) $\frac{5}{11}$

16. Coach-father-coach, see answer to 17 of Exercises 6.1.

17. See solutions to the prisoner's dilemma of problem 38.

18. $Pr[E] = \frac{_4C_2}{2^4} = 3/8$, $Pr[F] = 3/8$, $Pr[G] = \frac{_4C_1 \times 2}{2^4} = 1/2$

19. (a) The number of ways three passengers can be chosen for the first car is $_9C_3$ and the number of ways the remaining six passengers can sit in the other two cars is 2^6.

Hence $Pr[E] = \frac{_9C_3 \times 2^6}{3^9} = \frac{1792}{6561}$

(b) $Pr[F] = \frac{_9C_3 \times _6C_3}{3^9} = \frac{560}{6561}$.

(c) $Pr[G] = \frac{_9C_2 \times _7C_3 \times 3!}{3^9} = \frac{7560}{3^9}$

20. (a) 1/70   (b) 16/70   (c) 16/70   (d) 36/70

21. The second player can win the runner-up if and only if he does not have to meet the best player before the finals, which can happen if and only if he is seeded not on the same half as the best player. Hence the required probabilty is $\frac{4}{7}$.

22. Note that if an independent trial with probability p is repeated n times then the expected number of success is np. Therefore, if the expected number is 1, then

n = $\frac{1}{p}$. After getting a number in the first box, the probability of getting a new

number is $\frac{4}{5}$ and the number of boxes needed to get a new number is $\frac{1}{4/5} = \frac{5}{4}$.

Similarly, after getting two different numbers, the probabilty of getting a new number is 3/5 and the number of boxes needed to obtain this is 5/3, the number of boxes needed for the fourth number is 5/2 and the number of boxes needed for the last number is 5. Hence the expected number of boxes = 1 + 5/4 + 5/3 + 5/2 + 5

$$= 6 + 5 \times \frac{13}{12} = 11.42 \text{ (approx.).]}$$

24. If the first ball is red, then no matter which basket it is drawn from, the baskets would be left with the same number of red balls as the number of black balls. So, the second ball provides no clue. Therefore, if a red ball is drawn first, put it back. If a black ball is drawn first do not put it back. Then figure out the probability in each possible case and identify the one that has the higher probability as shown in the tree diagram below:

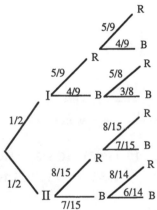

Outcome	Probability from I	Probability from II	Decision
RR	$\frac{5}{9} \times \frac{5}{9} = .31$	$\frac{8}{15} \times \frac{8}{15} = .28$	Basket I
RB	$\frac{5}{9} \times \frac{4}{9} = .246$	$\frac{8}{15} \times \frac{7}{15} = .248$	Basket II
BR	$\frac{4}{9} \times \frac{5}{8} = .27$	$\frac{7}{15} \times \frac{8}{14} = .26$	Basket I
BB	$\frac{4}{9} \times \frac{3}{8} = .17$	$\frac{7}{15} \times \frac{6}{14} = .2$	Basket II

25. $\frac{1}{2} \times [.31 + .248 + .27 + .2] = .514$

27. $\frac{1}{2} \times [\frac{4}{9} + \frac{12}{49} + \frac{1}{3} + \frac{1}{7}] = .58$

28. A should get $(\frac{7}{8})$th of the stakes and B should get $(\frac{1}{8})$th. This is arrived at by figuring out the probability. Since it is a fair game, the probability of A or B winning is 1/2. The following tree diagram shows that Pr[B winning] = 1/8 and Pr[A winning] = 7/8.

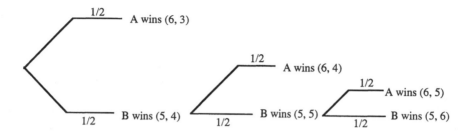

29.   A gets $\frac{3}{4}$ th and B gets $\frac{1}{4}$ th.

30. (a) Pr[6 appears at least once] = 1 - $(5/6)^{24}$ = 0.518, Odds in favor 518 : 482

    (b) Pr[6 appears at least once] = 1 - $(35/36)^{24}$ = 0.4914, odds in favor are 4914 : 5086.

31. Whichever hand gets the first ace, the other ace can be one of the twelve remaining cards wheras it can be in the other hand one of the thirteen cards. So, the probability is $\frac{12}{25}$.

32. This is like the previous problem, where the split was 2 - 0. In this case the split is 3 - 0 for the selected cards. We can use Zweifel's formula, [P. W. Zweifel, *Some Remarks About Bridge Probabilities*, Math Mag, Vol .59, No.3, 1986, 153 - 157] for calculating the probability of an m - n split:

$$\frac{1}{2} \times K \times \frac{12}{25} \cdot \frac{11}{24} \cdots \frac{13 - (m - 1)}{25 - (m -1)} \cdot \frac{13}{26 - m} \cdots \frac{13 - (n - 1)}{26 - (m + n - 1)},$$

33. (a) 4/9   (b) $\dfrac{4}{9} \times \dfrac{3}{8} = \dfrac{1}{6}$ (c) This is a 2 - 2 split with the number of cards 10.

Therefore, modifying Zwifel's formula we get $\dfrac{1}{2} \times {}_4C_2 \times \dfrac{4}{9} \times \dfrac{5}{8} \times \dfrac{4}{7} = \dfrac{10}{21}.$

36. Rabbi Ben Ezra gave two solutions:

Solution 1: Divide the estate proportionately. Since $1 + (1/2) + (1/3) + (1/4) = 25/12.$
Divide the estate by 25/12 so that

Reuben gets $\dfrac{1}{(25/12)} = \dfrac{12}{25}$,   Simeon receives $\dfrac{1}{2} \times \dfrac{12}{25} = \dfrac{6}{25}$

Levi gets $\dfrac{1}{3} \times \dfrac{12}{25} = \dfrac{4}{25}$,   Judah receives $\dfrac{1}{4} \times \dfrac{12}{25} = \dfrac{3}{25}.$

Rabbi Ben Ezra stated this to be the mathematician's solution.

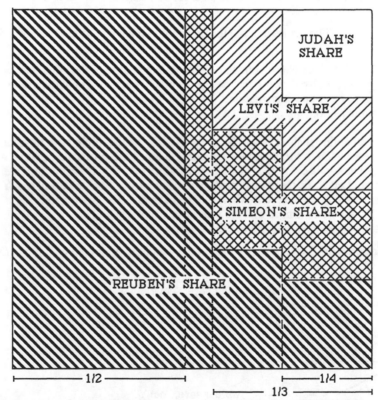

Solution 2: Divide the estate in proportion to the expected value of each claim. This is
done in the following manner. Divide the estate into four parts: part I $= \dfrac{1}{4}$ where
each of the four brothers has a claim and everybody has equal claim on it so

everybody gets (1/4)th, on part II $= \frac{1}{3} - \frac{1}{4}$ Reuben, Simeon Levi have equal claim so each of them gets $\frac{1}{3}$ of this, part III $= \frac{1}{2} - \frac{1}{3}$ is claimed by Reuben and Simeon and each of them get $\frac{1}{2}$ of this and part IV $= \frac{1}{2}$ which is claimed by only Reuben and it goes to him. [See diagram above].
Rabbi Ben Ezra stated this to be the Rabbis' solution.

37.   (a)   1/1,000   (b)   1/1,000,000   (c)   1/1,000

38.   (a) $\frac{1}{10^4}$ (b) $\frac{1}{10^8}$ (c) $\frac{1}{10^4}$

39.   The probability of winning is 1/3 for the hippo, 2/5 for the rhinoceros. Since the total probablities of the three must add to 1, the probability of winning for the giraffe is $1 - (1/3) - (2/5) = 4/15$. So the odds are 11 : 5 against the giraffe.

40.   Fallacious. If he carries a bomb the probability of somebody else bombing it is still $\frac{1}{10^6}$.

41.   (a) 1/3. No. You chose your card thinking that neither of the other two cards is the queen. So, by showing that one of the other two cards is not the queen, he is not giving you any new information.
     (b) same as (a).

42   The probability that Mr X will be hanged is 1/3. So, he might have thought that if he knew which of the other two will not be hanged, the probability that he would be hanged would become 1/2, which is larger than 1/3. However, this argument is not correct. Let X be the event Mr. X will be hanged, Y be the event Mr. Y will be hanged and Z be the event Mr. Z will be hanged. Then (see diagram)
     Pr[X will be hanged and the warder says 'not Y'] = 1/6
Pr[the warder says 'not Y'] = 1/6 + 1/3 = 1/2

Therefore, Pr[X will be hanged given the warder says 'not Y'] $= \dfrac{1/6}{1/2} = 1/3$.

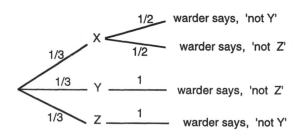

43.   In this case, the observations are the same as in Table 1, however, some ingenuity is required to determine how to distribute the marbles for the two weighings. For

simplicity, let us designate the marbles as R_1, R_2, W_1, W_2, B_1, B_2, where R, W, B stands for red, white and black respectively. First, let us answer a few questions:

A. Should we weigh one against one? The answer is no, because it will require three weighings.

B. Should we weigh three against three? The answer is no because that way, we get only two possibilities ↓ or ↑ (= cannot occur since there are three heavy ones).

C. So the obvious choice is to weigh two against two. We, however, need to make sure that each of the marble is weighed at least once.

Accordingly, the weighings should be as follows:

Weighing	Left pan	Right pan
First	$R_1 W_1$	$R_2 B_2$
Second	$B_1 W_2$	$W_1 R_2$

This time note that in the first weighing:

If ↓ then possibly heavy marbles are: $R_1 W_1 B_1$ or $R_1 W_1 B_2$ or $R_1 W_2 B_1$.

If ↑ then possibly heavy marbles are: $R_2 W_1 B_2$ or $R_2 W_2 B_1$ or $R_2 W_2 B_2$.

If = then possibly heavy marbles are: $R_1 W_2 B_2$ or $R_2 W_1 B_1$.

Similarly, in the second weighing:

If ↓ then possibly heavy marbles are: $R_1 W_2 B_1$ or $R_2 W_2 B_1$ or $R_1 W_2 B_2$.

If ↑ then possibly heavy marbles are: $R_2 W_1 B_1$ or $R_2 W_1 B_2$ or $R_1 W_1 B_2$.

If = then possibly heavy marbles are: $R_2 W_2 B_2$ or $R_1 W_1 B_1$.

From the above observations, we can identify the heavy marbles as follows:

If	then
↓↓	$R_1 W_2 B_1$
↓↑	$R_1 W_1 B_2$
↓ =	$R_1 W_1 B_1$
↑↓	$R_2 W_2 B_1$
↑↑	$R_2 W_1 B_2$
↑ =	$R_2 W_2 B_2$
= ↓	$R_1 W_2 B_2$
= ↑	$R_2 W_1 B_1$

= =   cannot occur.

44. (a) Every element of S is either in a subset of S or not. So, in forming a subset there are 2 possibilities for each element of S. Since there are n elements in S, the total number of possibilities is 2^n.

    (b) For any element $a \in S$, consider the set $S - \{a\}$. The set $S - \{a\}$ has n - 1 elements, so it has 2^{n-1} subsets. None of the subsets of $S - \{a\}$ has 'a' as an element. By adding 'a' as an element to each subset of $S - \{a\}$, we get all the subsets of S containing 'a'. The total number of the subsets of S containing 'a' will therefore be 2^{n-1}.

    (c) For every subset A in the collection S', the complement of A in S cannot be a member of S'. Hence S' can have at most half of the possible subsets of S. Therefore, S' can have at most $2^n \div 2 = 2^{n-1}$ members.

45. A bridge hand consists of 13 cards. So, $n(U) = {}_{52}C_{13} = 635{,}013{,}559{,}600$. The number of ways of getting 5 spades out of 13 is ${}_{13}C_5$. Similarly, the number of ways of getting 4 hearts is ${}_{13}C_4$, the number of ways of getting 3 diamonds is ${}_{13}C_3$ and that of 1 club is ${}_{13}C_1$. So, the rquired probability is $\dfrac{{}_{13}C_5 \times {}_{13}C_4 \times {}_{13}C_3 \times {}_{13}C_1}{{}_{52}C_{13}}$.

Chapter 11, Exercises 11.1

1. This can be done by tossing a coin. Designate the outcome head as boy and the outcome tail as girl. Toss the coin three times and note the outcome and repeat it a fairly large number of times and find the frequency of each outcome.

2, 3. Similar to 1.

4. Write each of the numbers 1 through 5 on a piece of paper, fold them and put in a hat. After mixing them draw one number. Repeat this till each number turns up at least once. Note the number of draws. Then repeat the process a large number of times and use the mean number of draws to answer the question.

5. Use nine identical cards from two suits, say $A\diamond, 2\diamond, 3\diamond, 4\diamond, 5\diamond, 6\diamond, 7\diamond, 8\diamond, 9\diamond$ and $A\spadesuit, 2\spadesuit, 3\spadesuit, 4\spadesuit, 5\spadesuit, 6\spadesuit, 7\spadesuit, 8\spadesuit, 9\spadesuit$. We designate each diamond card as one of the players and the corresponding spade card as his hat. Shuffle the cards separately and put them in two different piles. Then draw one card from each pile, if the cards match then it is a case of a player getting his own hat otherwise it is a case of a player getting somebody else's hat. Repeat this experiment and find the mean.

6. Take nine cards of the same suit, shuffle and draw one card. Put the card back and repeat till each card is drawn at least once. Repeat a large number of times and use the mean number of draws to answer the question.

Chapter 11, Exercises 11.2

1. (a) $a^2 - 2ab + b^2$ (b) $a^3 - 3a^2b + 3ab^2 - b^3$
    (c) $a^5 - 5a^4b + 10a^3b^2 - 10a^2b^3 + 5ab^4 - b^5$
    (d) $a^8 - 8a^7b + 28a^6b^2 - 56a^5b^3 + 70a^4b^4 - 56a^3b^5 + 28a^2b^6 - 8ab^7 + b^8$

2. (a) $x^2 + 2xy + y^2$ (b) $x^2 + 4xy + 4y^2$ (c) $9x^2 - 12xy + 4y^2$

3. (a) $1 + nx + nC_2x^2 + \ldots + x^n$
    (b) $1 - nx + nC_2x^2 - \ldots + (-1)^nx^n$
    (c) $(100 + 2)^2 = 1,061,208$
    (d) $(100 + 10)^4 = 146,410,000$
    (e) $(100 - 1)^2 = 9801$
    (f) $(1000 - 1)^3 = 997,002,999$
    (g) $1 - n(1/x)^2 + \ldots + (-1)^n(1/x)^n$
    (h) $a^3 + 6a^2b + 12ab^2 + 8b^3$
    (i) $a^3 - 6a^2b + 12ab^2 - 8b^3$
    (j) $81a^4 - 108a^3b + 54a^2b^2 - 12ab^3 + b^4$

4. $11^2 = 121$, $11^3 = 1331$, $11^4 = 14641, \ldots, 11^n =$ the number formed by the coefficients in the n-th row of Pascal's triangle. [see answer to 5.]

5. From the binomial theorem, we get

    $$11^n = (10 + 1)^n = 10^n + n(10)^{n-1} + \ldots + nC_2 \times 10^2 + n(10) + 1$$

    Therefore, 11^n has 1 unit, n tens, nC_2 hundreds, nC_3 thousands, and so on. So,

    $$11^7 = 19487171$$

    since it has 1 unit, 7 tens, 21 hundreds, 35 thousands, 35 ten thousands, and so on.

6. 1, 7, 21, 35, 35, 21, 7, 1; So,
    $(a + b)^7 = a^7 + 7a^6b + 21a^5b^2 + 35\ a^4b^3 + 35\ a^3b^4 + 21a^2b^5 + 7ab^6 + b^7$

7. (a) $2^n = (1 + 1)^n = 1 + n + nC_2 + \ldots + n + 1$
    (b) given expression $= n[1 + (n - 1) + \dfrac{(n-1)(n-2)}{2} + \ldots + 1]$
    $= n(1 + 1)^{n-1}\ \ = n(2)^{n-1}$

INDEX

A

Alternate hypothesis, 241
Arithmetic mean, 36
Astragalus, 159

B

Bar graph, 4
Batting average, 31
Bernoulli, 260
Bills of Mortality, 261
Binomial
    coefficients, 258
    experiment, 157
    theorem, 164, 256
Binomial distribution, 164
    graph of, 174
    standard deviation of, 166
Binomial probability, table for, 160

C

Caldwell, Glenn, 170
Cardan, 260
Cards, 74
Central limit theorem, 181, 229
Central tendency, 36
Chebyshev, 208, 261
Chebyshev's
    inequality, 208
    theorem, 208
Chuck-a-luck, 196
Circle graphs, 11
Combinations, 106
Complements, 79
Conditional probability, 135
Confidence
    interval, 231
    level, 235
    limits, 235
Cooper, Paul, 170
Correlation, 58
    coefficient, 59
    direct, 59
    high, 59
    inverse, 59, 61
    perfect, 59

Counting
    Fundamental Principle of, 100
    techniques, 96
    with sets, 89
CRAPS, 247
Critical ratio, 242
Cumulative frequency, 24
Cumulative frequency polygon, 25

D

Data,
    bimodal, 16, 18
    grouped, 46
    organization of, 1
    qualitative, 4
    quantitative, 4
    raw, 3
    ranked, 27
    skewed, 16, 18, 42
        negatively, 42
        positively, 42
    symmetric, 16
Deadly occupations, 9, 11
De Mere's problem, 249, 259
De Moivre, 260, 265
Deviation, 49
Dewey, 263
Disjoint sets, 80
Dispersions, 49
Duke of Tuscany, 124, 259

E

Earned run average, 35
Empty set, 77, 84
Equiprobable measure, 120
ERA, 35
Estimating binomial probability, 185
Estimating population
    proportion, 230
    size, 237
Event, 114
    independent, 136
    mutually exclusive, 125
    probability measure of, 116

Expected
    loss, 132
    number of successes, 162
    value of a game, 193
    win, 132
Experiment, 96

F
Face cards, 74
Factorial notation, 104
Fair bet, 193
Favorable game, 193
Fermat, 259
Finite set, 87
Finite stochastic process, 146
First success
    number of trials for, 203
Fisher, Ronald, 261
Fra Luca Paccioli's problem, 249
Frequency distribution, 1
    graph of, 41
    with grouped data, 20
    negatively skewed, 42
    polygon, 22
    positively skewed, 42
    skewed, 41
    symmetrical, 41
    table, 3

G
Galileo, 259
Gallup Poll, 220
Gallup, George H., 220, 225, 263
Galton, Francis, 262
Gore, Senator, 227
Graunt, John, 261

H
Halley, George, 262
Harris, Louis, 264
Harris Poll, 264
Helen of Troy, 246
Hilbert, 261
Histogram, 21
Honor cards, 74
Huygens, 260

I
Independent
    events, 136
    trials with two outcomes, 155
Inductive Statistics, 214

Infinite sets, 87
Interval, length of, 20

L
Laplace, 261
Law of large numbers, 173, 260
Line chart, 4
Loyds, Sam, puzzle, 251
Loaded die, 117

M
Margin of error, 226, 231, 236
Markov, 261
Mean , 36, 40
    deviation, 41
    of a binomial distribution, 164
    of grouped data, 46
    of a random variable, 193
Measures of Variation, 49
Median, 27, 40
Modal class, 47
Mode, 3, 40
Moesteller, 245
Monte Carlo methods, 253, 262
Morra, 246
Mothers' smoking, 39
Multiplication vs. addition, 101
Mutually exclusive events, 125

N
National Center for Health Statistics, 14
Natural numbers, 88
Negatively skewed, 42
Neumann, 263
Normal Curve, 180
    table for areas under, 182
Null hypothesis, 241
Number of elements, 70

O
Odds against, 132
Odds in favor, 131
One-to-one correspondence, 87
Operation Smoky, 172
Opinion poll, 220, 263

P
Pascal, 256, 264
Pascal's triangle, 256
Pearson, Karl, 262

Percentiles, 28
Permutations, 103
Permutations vs. Combinations, 109
Pie charts, 11
    size of, 13
Poll by phone, 223
Population, 215
    control, 64
    estimating size of, 237
Positively skewed data, 42
Principles of disclosure, 222
Prisoner's dilemma, 252
Probability,
    a priori, 214
    estimates, 234
    measure, 114, 116, 125
Product of deviations, 59

Q
Quartiles, 27
Quetelet, Lambert, 262

R
Rabbi Ben Ezra, 250, 259
Random
    devices, 253
    digits, 217
    phenomena, 129
    sample, 215
Random variable, 192
    expected value of, 193
    probability distribution of, 198
    standard deviation of, 206
Range, 49
Relative frequency, 7

S
Sample, 215
    biased, 215
    fair, 215
    size of, 224
Sample space, 96
Selected at random, 120
Sets
    equality of, 85
    finite, 87
    infinite, 87
    intersection of, 69, 79
    union of, 67
    universal, 69
    well-defined, 76

Set builder notation, 79
Sigma notation, 36
Slugging average, 31
Standard deviation, 53
    of binomial distribution, 166
    of a random variable, 206
Stem and leaf display, 16
Stochastic process, 146
Subset, 84
    proper, 84
    improper, 84
Symmetric data, 16
Symmetrical distribution, 41

T
Test of hypothesis, 239
Tree diagram, 97
Truman, President, 263

U
Ulam, S.M., 262
Unfavorable game, 193
Universal sets, 67, 78

V
Vaccine problem, 168
Variables, 58
Variance, 52
    of binomial distribution, 166
Venn diagram, 68

W
Weights, 116

Y
Yaglom and Yaglom, 245

Z
z-value, 177
Zweifel's formula , 298
Zweifel, P. W., 298